Operator Theory - Recent Advances, New Perspectives and Applications

Edited by Abdo Abou Jaoudé

Published in London, United Kingdom

Operator Theory - Recent Advances, New Perspectives and Applications
http://dx.doi.org/10.5772/intechopen.102221
Edited by Abdo Abou Jaoudé

Contributors
Abdo Abou Jaoudé, Hanni Dridi, Fethi Ahmed Soltani, A. Torres-Hernandez, F. Brambila-Paz, R. Ramirez-Melendez, Jonathan Bartlett, Maria Isabelle Fite, Abdourahman Haman Adji

First published in London, United Kingdom, 2023 by IntechOpen
IntechOpen is the global imprint of INTECHOPEN LIMITED, registered in England and Wales, registration number: 11086078, 5 Princes Gate Court, London, SW7 2QJ, United Kingdom

British Library Cataloguing-in-Publication Data
A catalogue record for this book is available from the British Library

Additional hard and PDF copies can be obtained from orders@intechopen.com

Operator Theory - Recent Advances, New Perspectives and Applications
Edited by Abdo Abou Jaoudé
p. cm.
Print ISBN 978-1-83880-992-8
Online ISBN 978-1-83881-111-2
eBook (PDF) ISBN 978-1-83881-112-9

Meet the editor

Abdo Abou Jaoudé has been teaching for many years and has a passion for researching mathematics. He is currently an Associate Professor of Mathematics and Statistics at Notre Dame University-Louaizé (NDU), Lebanon. He holds a BSc and an MSc in Computer Science from NDU, and three Ph.D.'s in Applied Mathematics, Computer Science, and Applied Statistics and Probability, all from Bircham International University, Spain. He also holds two Ph.D.'s in Mathematics and Prognostics from the Lebanese University, Lebanon, and Aix-Marseille University, France. Dr. Abou-Jaoudé's broad research interests are in the fields of pure and applied mathematics, and he has published twenty-three international journal articles and six contributions to conference proceedings, in addition to ten books on prognostics, mathematics, and computer science.

Contents

Preface

It gives me great pleasure to introduce as well as discuss, learn, solve, teach, and work with operator theory. This book discusses some fundamental aspects of the theory of operators and explores their use to solve a large array of problems. The book addresses many topics, including the paradigm of complex probability, principal parts extension for a Noether operator A, the sets of fractional operators and some of their applications, the total and partial differentials as algebraically manipulable entities, and more.

In mathematics, operator theory is the study of linear operators on function spaces, beginning with differential operators and integral operators. The operators may be presented abstractly by their characteristics, such as bounded linear operators or closed operators; consideration may also be given to nonlinear operators. The study of operator theory, which depends heavily on the topology of function spaces, is a branch of functional analysis. Also, if a collection of operators forms an algebra over a field, then it is an operator algebra. Hence, the description of operator algebras is part of operator theory.

Moreover, single-operator theory deals with the properties and classification of operators, considered one at a time. For example, the classification of normal operators in terms of their spectra falls into this category. In addition, the theory of operator algebras brings algebras of operators such as C^*-algebras to the fore.

Many operators that are studied are operators on Hilbert spaces of holomorphic functions, and the study of the operator is intimately linked to questions in function theory. For example, Beurling's theorem describes the invariant subspaces of the unilateral shift in terms of inner functions, which are bounded holomorphic functions on the unit disk with unimodular boundary values almost everywhere on the circle. Beurling interpreted the unilateral shift as multiplication by the independent variable on the Hardy space. The success in studying multiplication operators, and more generally Toeplitz operators (which are multiplication, followed by projection onto the Hardy space), has inspired the study of similar questions in other spaces, such as the Bergman space. Therefore, operator theory has a connection with complex analysis.

This volume illustrates the use of operator theory when applied to solve specific problems and discusses some fundamental aspects of pure and applied operator theory. It is a useful resource for scholars, researchers, and undergraduate and graduate students in pure and applied mathematics, classical and modern physics, engineering, and science in general.

Abdo Abou Jaoudé, Ph.D.
Associate Professor,
Mathematics and Statistics,
Notre Dame University-Louaizé,
Zouk Mosbeh, Lebanon

Chapter 1

The Paradigm of Complex Probability and the Theory of Metarelativity: A Simplified Model of MCPP

Abdo Abou Jaoudé

"Subtle is the Lord. Malicious, He is not."

Albert Einstein.

"Mathematics, rightly viewed, possesses not only truth but supreme beauty… "

Bertrand Russell.

"Logic will get you from A to Z; imagination will get you everywhere."

Albert Einstein.

"There are more things in heaven and earth, Horatio, than are dreamt of in your philosophy."

Hamlet (1601), William Shakespeare.

"Ex nihilo nihil fit: Nothing comes from nothing."

Parmenides.

"All is One. From One all things."

Heraclitus.

Abstract

All our work in classical probability theory is to compute probabilities. The original idea in this research work is to add new dimensions to our random experiment, which will make the work deterministic. In fact, probability theory is a nondeterministic theory by nature; which means that the outcome of the events is due to chance and luck. By adding new dimensions to the event in the real set of probabilities $\mathcal{R}$, we make the work deterministic, and hence a random experiment will have a certain outcome in the complex set of probabilities and total universe $\mathbf{G} = \mathcal{C}$. It is of great importance that the stochastic system, like in real-world problems, becomes totally predictable since we will be totally knowledgeable to foretell the outcome of chaotic and random events that occur in nature, for example, in statistical mechanics or in all stochastic processes. Therefore, the work that should be done is to add to the real set of probabilities $\mathcal{R}$ the contributions of $\mathcal{M}$, which is the imaginary set of probabilities that will make the event in $\mathbf{G} = \mathcal{C} = \mathcal{R} + \mathcal{M}$ deterministic. If this is found to be fruitful, then a new theory in statistical sciences and in science, in general, is elaborated and this is to understand absolutely deterministically those phenomena that used to be random phenomena in $\mathcal{R}$. This paradigm was initiated and developed in my previous 21 publications. Moreover, this model will be related to my theory of Metarelativity, which takes into account faster-than-light matter and energy. This is what I called "The Metarelativistic Complex Probability Paradigm (*MCPP*)," which will be elaborated on in the present two chapters 1 and 2.

Keywords: chaotic factor, degree of our knowledge, complex random vector, probability norm, complex probability set $\mathcal{C}$, metarelativistic transformations, imaginary number, imaginary dimensions, superluminal velocities, metaparticles, dark matter, metamatter, dark energy, metaenergy, metaentropy, universe $\mathbf{G}_1$, metauniverse $\mathbf{G}_2$, luminal universe $\mathbf{G}_3$, the total universe $\mathbf{G}$

1. Introduction

The development of a new theory in physics, which I called the theory of Metarelativity creates a new continuum or space-time in which a new matter interacts [1–3]. This newly discovered matter is surely not the ordinary matter but a new kind of matter that can be easily identified to be the dark matter that astronomers, astrophysicists, and cosmologists seek to find. In fact, my novel theory shows that this new matter is superluminal by nature and is related to the new meta-space-time that lies in the metauniverse $\mathbf{G}_2$ in the same fashion that ordinary matter is related to the ordinary space-time and that lies in the universe $\mathbf{G}_1$ that we know. From what has been proved in Metarelativity, it was shown that the theory does not destroy Albert Einstein's theory of relativity that we know at all but on the contrary, it proves its veracity and expands it to the superluminal velocities' realm. The new space-time is "imaginary" since it exists in the domain of imaginary numbers and is now called meta-space-time or metauniverse because it lays beyond the ordinary "real" space-time that exists in the domain of real numbers as well as the matter and the energy interacting within them. Now the relation between both matter and metamatter is shown in the theory of Metarelativity. The first space-time is called the universe and the second space-time is called the metauniverse, which is another universe if we can say as material and as real as the first one but at a different level of experience because it is superluminal relative to the first one. It is similar to the atomic world that exists and is real but at a different level of physical experience, in the sense that we have discovered its laws in the theory of quantum mechanics where we deal with atoms and particles like when we deal in astronomy and astrophysics with planets and galaxies. In fact, astronomy is also real in the sense that we have discovered the laws governing the stars and planets but it lays at a different level of reality from our everyday world and experience. Metarelativity comes now to enlarge once more the scope of our understanding to encompass a new level of physical reality.

Furthermore, my Metarelativity will be bonded to my Complex Probability Paradigm (*CPP*), which was developed in my 21 previous research works. In fact, the system of axioms for probability theory laid in 1933 by Andrey Nikolaevich Kolmogorov can be extended to encompass the imaginary set of numbers, and this by adding to his original five axioms an additional three axioms. Therefore, we create the complex probability set $\mathcal{C}$, which is the sum of the real set $\mathcal{R}$ with its corresponding real probability and the imaginary set $\mathcal{M}$ with its corresponding imaginary probability. Hence, all stochastic and random experiments are performed now in the complex set $\mathcal{C}$ instead of the real set $\mathcal{R}$. The objective is then to evaluate the complex probabilities by considering supplementary new imaginary dimensions to the event occurring in the "real" laboratory. Consequently, the corresponding probability in the whole set $\mathcal{C}$ is always equal to one and the outcome of all random experiments that follow any probability distribution in $\mathcal{R}$ is now predicted totally and absolutely in $\mathcal{C}$. Subsequently, it follows that chance and luck in $\mathcal{R}$ are replaced by total determinism in $\mathcal{C}$. Consequently, by subtracting the chaotic factor from the degree of our knowledge of the stochastic system, we evaluate the probability of any random phenomenon in $\mathcal{C}$. My innovative Metarelativistic Complex Probability Paradigm (*MCPP*) will be developed in this work in order to express all probabilistic phenomena completely deterministically in the total universe $\mathbf{G} = \mathcal{C} = \mathcal{R} + \mathcal{M} = \mathbf{G}_1 + \mathbf{G}_2 + \mathbf{G}_3$.

Finally, and to conclude, this research work is organized as follows: After the introduction in Section 1, the purpose and the advantages of the present work are presented in Section 2. Afterward, in Section 3, we will review and recapitulate the complex probability paradigm (*CPP*) with its original parameters and interpretation. In Section 4, a concise review of Metarelativity will be explained and summarized. Also, in Section 5, I will show the road and explain the steps that will lead us to the final *MCPP* theory, which will be developed in the subsequent sections. Therefore, in Section 6, and after extending Albert Einstein's relativity to the imaginary and complex sets, I will link my original theory of Metarelativity to my novel complex probability paradigm; hence, the first simplified model of *MCPP* will be developed. Finally, in Section 7, we will present the conclusion of the first chapter and then mention the list of references cited in the current research work. Moreover, in the second following chapter and in Section 1, a more general second model will be established. Furthermore, in Section 2, a wider third model will be presented. And in Section 3, the final and the most general model of *MCPP*, which takes into account the case of electromagnetic waves will be elaborated. Additionally, in Section 4, we will present some very important consequences of the *MCPP* paradigm. Finally, I conclude the work by doing a comprehensive summary in Section 5 and then present the list of references cited in the second research chapter.

2. The purpose and the advantages of the current publication

To summarize, the advantages and the purposes of this current work are to [4–24]:

1. Extend the theory of classical probability to encompass the complex numbers set; hence, to bond the theory of probability to the field of complex variables and analysis in mathematics. This mission was elaborated on and initiated in my earlier 21 papers.

2. Apply the novel probability axioms and *CPP* paradigm to Metarelativity and hence to bond my Metarelativity theory to my Complex Probability Paradigm and thus show that:

$$\begin{aligned}\mathcal{C} &= \mathcal{R} + \mathcal{M} \\ &= (R_1 + R_2 + R_3) + (M_1 + M_2 + M_3) \\ &= (R_1 + M_1) + (R_2 + M_2) + (R_3 + M_3) \\ &= \mathcal{C}_1 + \mathcal{C}_2 + \mathcal{C}_3 = \mathbf{G}_1 + \mathbf{G}_2 + \mathbf{G}_3 = \mathbf{G}.\end{aligned}$$

3. Show that all nondeterministic phenomena like in the problems considered here can be expressed deterministically in the complex probabilities set and total universe $\mathbf{G} = \mathcal{C}$.

4. Compute and quantify both the degree of our knowledge and the chaotic factor of the probability distributions and *MCPP* in the sets $\mathcal{R}$, $\mathcal{M}$, and $\mathcal{C}$.

5. Represent and show the graphs of the functions and parameters of the innovative paradigm related to Metarelativity.

6. Demonstrate that the classical concept of probability is permanently equal to one in the set of complex probabilities; hence, no randomness, no chaos, no ignorance, no uncertainty, no nondeterminism, no unpredictability, and no information loss or gain exist in:

$$\begin{aligned}\mathbf{G}\ (\text{complex matter and energy set}) &= \mathcal{C}\ (\text{complex probabilities set}) \\ &= \mathcal{R}\ (\text{real probabilities set}) + \mathcal{M}\ (\text{imaginary probabilities set}).\end{aligned}$$

7. Explain the existence of dark matter and dark energy that exist in $\mathbf{G}_2 \subset \mathbf{G}$.

8. Prepare to implement this creative model to other topics and problems in physics. These will be the job to be accomplished in my future research publications.

Concerning some applications of the novel-founded paradigm and as a future work, it can be applied to any nondeterministic phenomenon in science. And compared with existing literature, the major contribution of the current research work is to apply the innovative paradigm of *CPP* to Metarelativity and to express it completely deterministically as well as to determine the corresponding mass and energy of dark matter and dark energy.

The next figure displays the major purposes of the Metarelativistic Complex Probability Paradigm (*MCPP*) (**Figure 1**).

3. The complex probability paradigm

3.1 The original Andrey Nikolaevich Kolmogorov system of axioms

The simplicity of Kolmogorov's system of axioms may be surprising [4–24]. Let E be a collection of elements $\{E_1, E_2, \ldots\}$ called elementary events and let F be a set of subsets of E called random events [25–29]. The five axioms for a finite set E are:

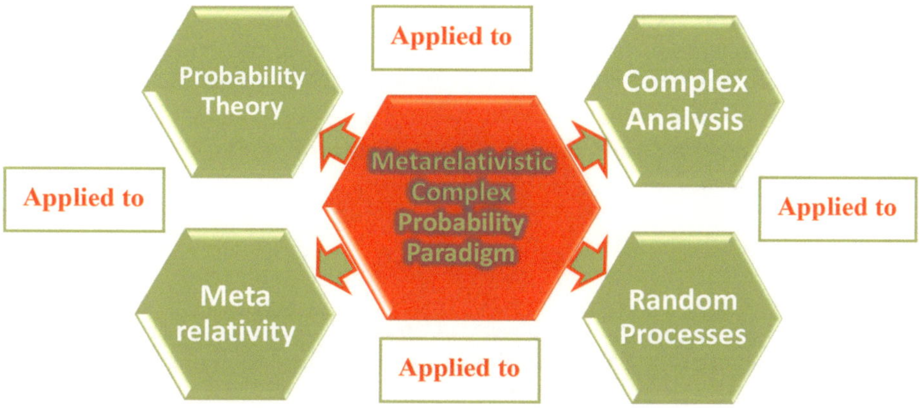

Figure 1.
The diagram of the Metarelativistic Complex Probability Paradigm's major purposes and goals.

Axiom 1: F is a field of sets.
Axiom 2: F contains the set E.
Axiom 3: A nonnegative real number $P_{rob}(A)$, called the probability of A, is assigned to each set A in F. We have always $0 \leq P_{rob}(A) \leq 1$.
Axiom 4: $P_{rob}(E)$ equals 1.
Axiom 5: If A and B have no elements in common, the number assigned to their union is:

$$P_{rob}(A \cup B) = P_{rob}(A) + P_{rob}(B)$$

Hence, we say that A and B are disjoint; otherwise, we have:

$$P_{rob}(A \cup B) = P_{rob}(A) + P_{rob}(B) - P_{rob}(A \cap B)$$

And we say also that: $P_{rob}(A \cap B) = P_{rob}(A) \times P_{rob}(B/A) = P_{rob}(B) \times P_{rob}(A/B)$ which is the conditional probability. If both A and B are independent then:

$$P_{rob}(A \cap B) = P_{rob}(A) \times P_{rob}(B).$$

Moreover, we can generalize and say that for N disjoint (mutually exclusive) events $A_1, A_2, \ldots, A_j, \ldots, A_N$ (for $1 \leq j \leq N$), we have the following additivity rule:

$$P_{rob}\left(\bigcup_{j=1}^{N} A_j\right) = \sum_{j=1}^{N} P_{rob}(A_j)$$

And we say also that for N independent events $A_1, A_2, \ldots, A_j, \ldots, A_N$ (for $1 \leq j \leq N$), we have the following product rule:

$$P_{rob}\left(\bigcap_{j=1}^{N} A_j\right) = \prod_{j=1}^{N} P_{rob}(A_j)$$

3.2 Adding the imaginary part $\mathcal{M}$

Now, we can add to this system of axioms an imaginary part such that:

Axiom 6: Let $P_m = i \times (1 - P_r)$ be the probability of an associated complementary event in $\mathcal{M}$ (the imaginary part or probability universe) to the event A in $\mathcal{R}$ (the real part or probability universe). It follows that $P_r + P_m/i = 1$, where i is the imaginary number with $i = \sqrt{-1}$ or $i^2 = -1$.

Axiom 7: We construct the complex number or vector $Z = P_r + P_m = P_r + i(1 - P_r)$ having a norm $|Z|$ such that:

$$|Z|^2 = P_r^2 + (P_m/i)^2.$$

Axiom 8: Let Pc denote the probability of an event in the complex probability set and universe $\mathcal{C}$, where $\mathcal{C} = \mathcal{R} + \mathcal{M}$. We say that Pc is the probability of an event A in $\mathcal{R}$ with its associated and complementary event in $\mathcal{M}$ such that:

$$Pc^2 = (P_r + P_m/i)^2 = |Z|^2 - 2iP_rP_m \text{ and is always equal to 1.}$$

We can see that by taking into consideration the set of imaginary probabilities we added three new and original axioms and consequently the system of axioms defined by Kolmogorov was hence expanded to encompass the set of imaginary numbers and realm.

3.3 A concise interpretation of the original *CPP* paradigm

To summarize the novel *CPP* paradigm, we state that in the real probability universe $\mathcal{R}$ the degree of our certain knowledge is undesirably imperfect and hence unsatisfactory, thus we extend our analysis to the set of complex numbers $\mathcal{C}$, which incorporates the contributions of both the set of real probabilities, which is $\mathcal{R}$ and the complementary set of imaginary probabilities, which is $\mathcal{M}$. Afterward, this will yield an absolute and perfect degree of our knowledge in the probability universe $\mathcal{C} = \mathcal{R} + \mathcal{M}$ because $Pc = 1$ constantly and permanently. As a matter of fact, the work in the universe $\mathcal{C}$ of complex probabilities gives way to a sure forecast of any stochastic experiment, since in $\mathcal{C}$ we remove and subtract from the computed degree of our knowledge the measured chaotic factor. This will generate in the universe $\mathcal{C}$ a probability equal to 1 as it is shown and proved in the following equation: $Pc^2 = DOK - Chf = DOK + MChf = 1 = Pc$. Many applications which take into consideration numerous continuous and discrete probability distributions in my 21 previous research papers confirm this hypothesis and innovative paradigm [4–24]. The Extended Kolmogorov Axioms (*EKA* for short) or the Complex Probability Paradigm (*CPP* for short) can be shown and summarized in the next illustration (**Figure 2**):

4. A concise review of Metarelativity

4.1 The new Metarelativistic transformations

Consider the following two inertial systems of referential (**Figure 3**) [1]:

Assume that v becomes greater than c, which is the velocity of light, the system of equations called metarelativistic transformations, which are the extension of relativistic Lorentz transformations to the imaginary space-time are:

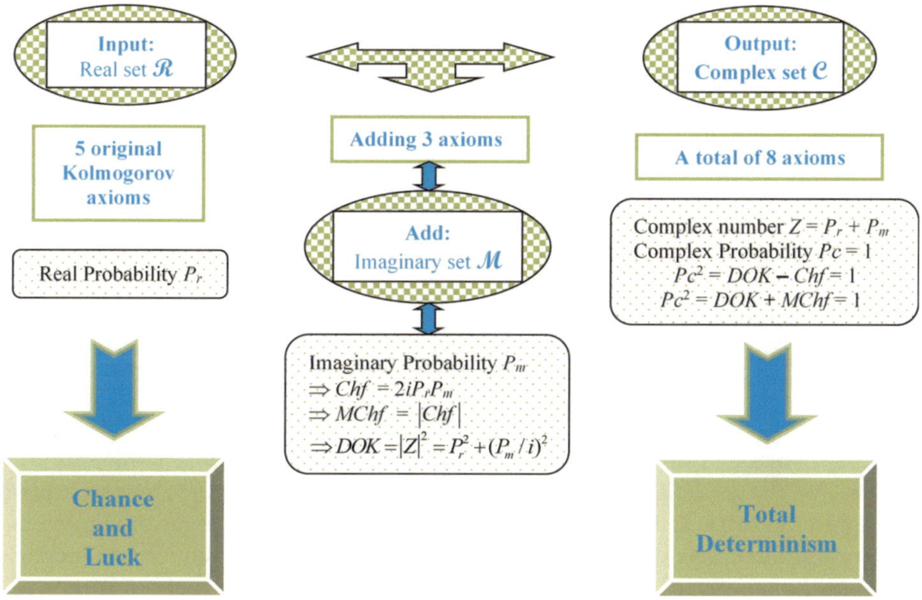

Figure 2.
The EKA *or the* CPP *diagram.*

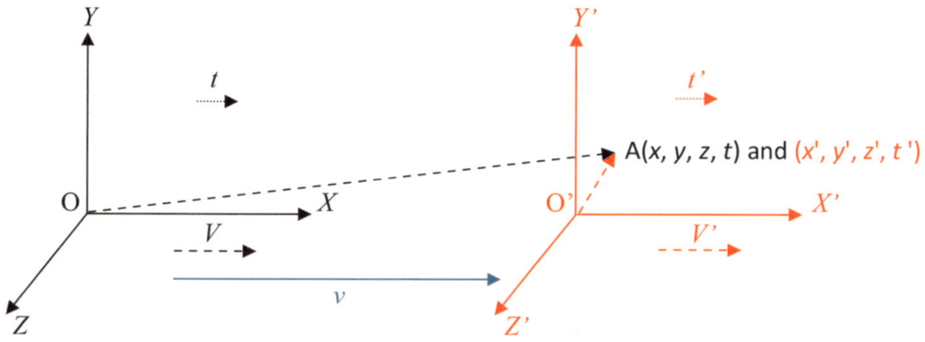

Figure 3.
Two inertial systems of referential.

$$\begin{cases} x' = \dfrac{(x - vt)}{\sqrt{(-1) \times \left(\dfrac{v^2}{c^2} - 1\right)}} = \dfrac{(x - vt)}{\sqrt{-1} \times \sqrt{\dfrac{v^2}{c^2} - 1}} = \dfrac{(x - vt)}{\sqrt{i^2} \times \sqrt{\dfrac{v^2}{c^2} - 1}} = \dfrac{(x - vt)}{\pm i \times \sqrt{\dfrac{v^2}{c^2} - 1}} = \dfrac{\pm i \times (x - vt)}{\sqrt{\dfrac{v^2}{c^2} - 1}} \\ y' = y \\ z' = z \\ t' = \dfrac{\left(t - \dfrac{vx}{c^2}\right)}{\sqrt{(-1) \times \left(\dfrac{v^2}{c^2} - 1\right)}} = \dfrac{\left(t - \dfrac{vx}{c^2}\right)}{\sqrt{-1} \times \sqrt{\dfrac{v^2}{c^2} - 1}} = \dfrac{\left(t - \dfrac{vx}{c^2}\right)}{\sqrt{i^2} \times \sqrt{\dfrac{v^2}{c^2} - 1}} = \dfrac{\left(t - \dfrac{vx}{c^2}\right)}{\pm i \times \sqrt{\dfrac{v^2}{c^2} - 1}} = \dfrac{\pm i \times \left(t - \dfrac{vx}{c^2}\right)}{\sqrt{\dfrac{v^2}{c^2} - 1}} \end{cases}$$

where i is the imaginary number such that: $i^2 = -1$ and $\sqrt{-1} = \pm i$ and $\frac{1}{\pm i} = \mp i$.

4.2 The mass of matter and metamatter

We have in special relativity: $m = \frac{m_0}{\sqrt{1-\frac{v^2}{c^2}}} = m_{G1}$.

If v becomes greater than $c = 300,000$ km/s, then m becomes equal to:

$$m = \frac{m_0}{\sqrt{(-1) \times \left(\frac{v^2}{c^2} - 1\right)}} = \frac{m_0}{\sqrt{-1} \times \sqrt{\frac{v^2}{c^2} - 1}} = \frac{m_0}{\sqrt{i^2} \times \sqrt{\frac{v^2}{c^2} - 1}} = \frac{m_0}{\pm i \times \sqrt{\frac{v^2}{c^2} - 1}}$$
$$= \frac{\pm i \times m_0}{\sqrt{\frac{v^2}{c^2} - 1}} = m_{G2}$$

yielding hence two imaginary and superluminal particles $+i|m_{G2}|$ and $-i|m_{G2}|$ where $|m_{G2}|$ is the module or the norm of the imaginary mass m_{G2}. They are called imaginary in the sense that they lay in the four-dimensional superluminal universe of imaginary numbers that we call $\mathbf{G_2}$ or the metauniverse. Matter which has increased throughout the whole process of special relativity will become equal to infinity when velocity reaches c as it is apparent in the equations, and in the new dimensions matter is imaginary due to the imaginary dimensions that we defined in the theory of Metarelativity. We say that beneath c we are working in the subluminal universe $\mathbf{G_1}$ or in the universe, and beyond c that we are working in $\mathbf{G_2}$ or in the metauniverse. Additionally, if the velocity is equal to c, we say that we are working in the luminal universe of electromagnetic waves and that is denoted by $\mathbf{G_3}$.

Firstly, in the first following equation:

$m = \frac{+im_0}{\sqrt{\frac{v^2}{c^2}-1}} = m_{G2}(\text{MetaParticle})$.

That means that matter now will decrease till it vanishes whenever the velocity reaches infinity, which means that mass is equal to zero at the velocity infinity.

Secondly, in the second following equation:

$m = \frac{-im_0}{\sqrt{\frac{v^2}{c^2}-1}} = m_{G2}(\text{MetaAntiParticle})$.

we say mathematically, that if v tends to infinity, m tends to zero. Matter now will continue increasing as in the equation till it vanishes whenever the velocity reaches infinity, which means that mass is equal to zero at the velocity infinity.

Thirdly, electromagnetic waves (EW), which travel at the velocity of light c exist in the universe $\mathbf{G_3}$ and have a mass:

$$m_{G3} = m_{EW} = \frac{hf}{c^2}$$

where h is Planck's constant and f is the frequency of the EW.

In the total universe $\mathbf{G} = \mathbf{G_1} + \mathbf{G_2} + \mathbf{G_3}$ we have:

$$m_G = m_{G1} + m_{G2} + m_{G3} = m_{G1} \pm i|m_{G2}| + \frac{hf}{c^2}$$

And we can notice that m_G belongs to the set of complex numbers denoted in mathematics by $\mathbb{C}$. The following graphs illustrate these facts (**Figures 4** and 5).

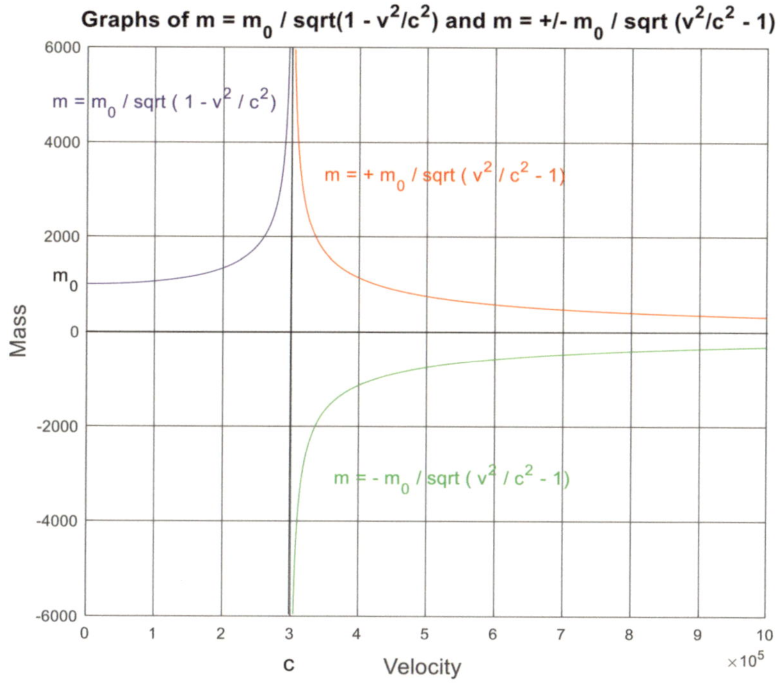

Figure 4.
The graphs of $m = m_0/\sqrt{1 - \frac{v^2}{c^2}}$ *in blue for* $0 \leq v < c$ *and of* $m = +m_0/\sqrt{\frac{v^2}{c^2} - 1}$ *in red and of* $m = -m_0/\sqrt{\frac{v^2}{c^2} - 1}$ *in green for* $v > c$.

Graphically, we can represent the two complementary metaparticles $+i|m_{G2}|$ and $-i|m_{G2}|$ and their annihilation in **Figure 6** as in the following reactions:

$+i|m_{G2}| - i|m_{G2}| \rightarrow 2m_{G1}$ and $+i|m_{G2}| - i|m_{G2}| \rightarrow 2$ photons.

4.3 The energy and the metaenergy

We know from special relativity that energy is given by:

$$E = mc^2 = \frac{m_0 c^2}{\sqrt{1 - \frac{v^2}{c^2}}} = E_{G1}$$

In Metarelativity, we have accordingly the imaginary energy or metaenergy. It is clear from the equation above that this metaenergy can be positive as:

$E = \frac{+im_0 \times c^2}{\sqrt{\frac{v^2}{c^2} - 1}} = E_{G2}$ (MetaParticle Energy or MetaEnergy).

or it can be negative as:

$E = \frac{-im_0 \times c^2}{\sqrt{\frac{v^2}{c^2} - 1}} = E_{G2}$(MetaAntiParticle Energy or MetaAntiEnergy).

Additionally, the luminal electromagnetic waves (EW) in the universe $\mathbf{G_3}$ have energy:

$$E_{G3} = E_{EW} = hf$$

where h is Planck's constant and f is the frequency of the EW.

Therefore, in the total universe $\mathbf{G} = \mathbf{G_1} + \mathbf{G_2} + \mathbf{G_3}$ we have:

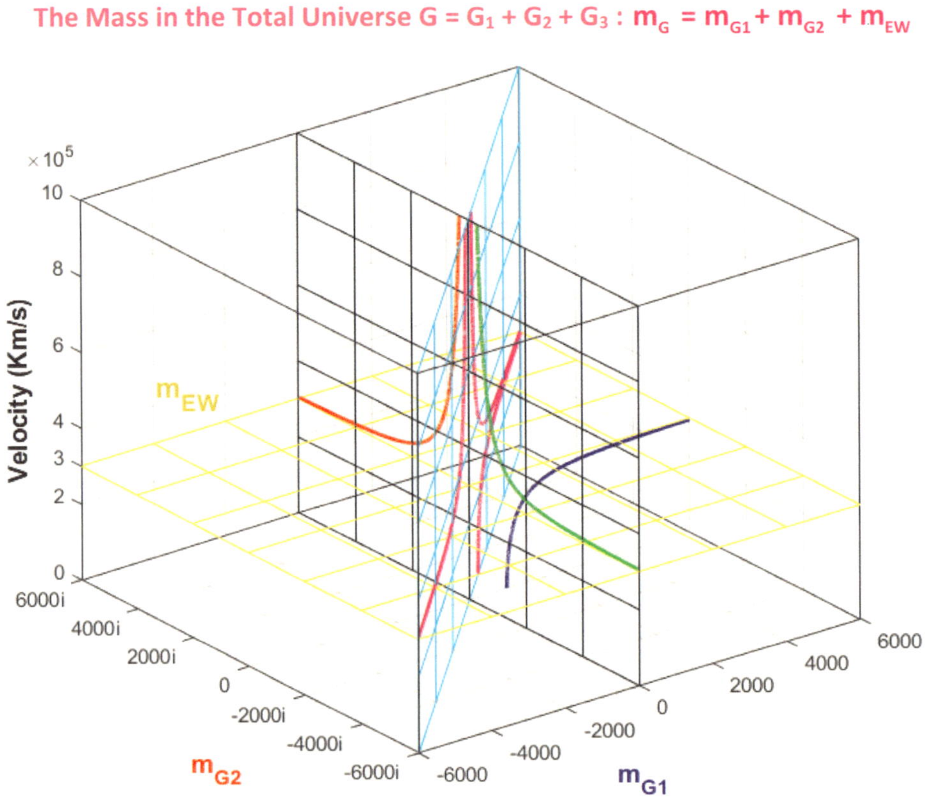

Figure 5.
The graphs of $m_{G1}(v)$ in blue, of $+i|m_{G2}(v)|$ in red, of $-i|m_{G2}(v)|$ in green, where m_{EW} lies in the yellow plane of equation v = c, and of $m_G = m_{G1} + m_{G2} + m_{EW} = m_{G1} \pm i|m_{G2}| + \frac{hf}{c^2}$ in magenta for $0 \leq v < +\infty$ in the complex plane in cyan.

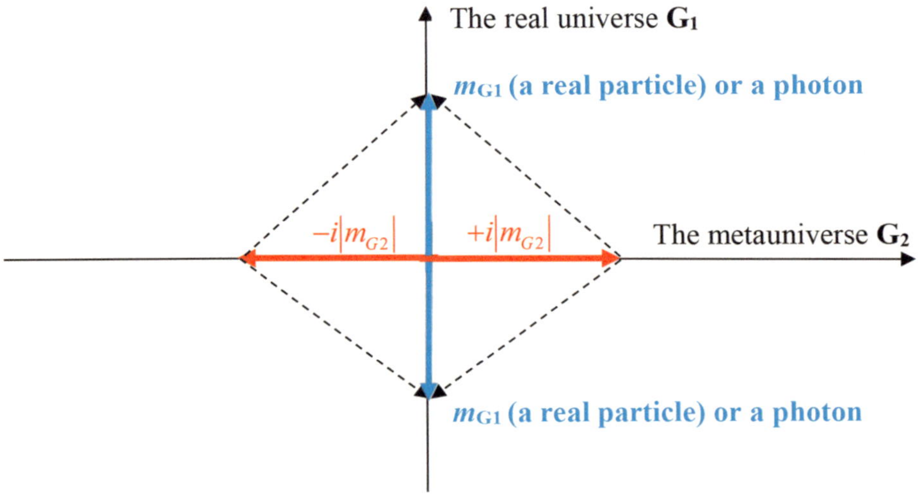

Figure 6.
The two complementary particles of metamatter $+i|m_{G2}|$ and $-i|m_{G2}|$ where $|m_{G2}|$ is the module or the norm of the imaginary mass m_{G2} in the metauniverse G_2 and their annihilation into the real matter in the universe G_1 or into photons in the universe G_3.

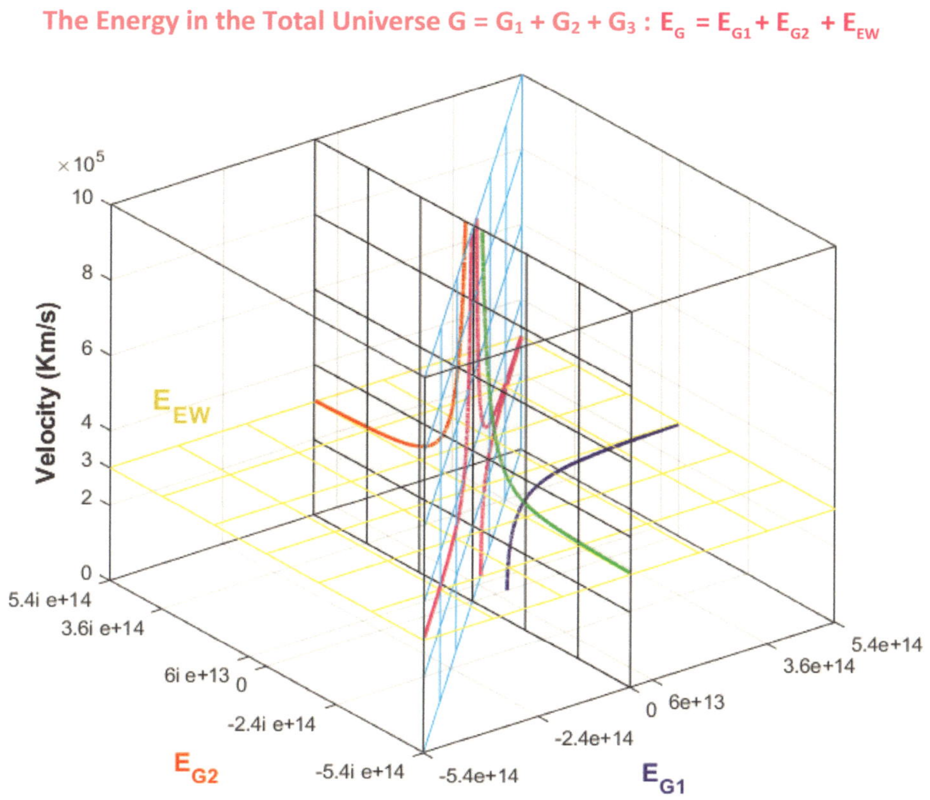

Figure 7.
The graphs of $E_{G1}(v)$ in blue, of $+i|E_{G2}(v)|$ in red, of $-i|E_{G2}(v)|$ in green, where E_{EW} lies in the yellow plane of equation v = c, and of $E_G = E_{G1} + E_{G2} + E_{EW} = E_{G1} \pm i|E_{G2}| + hf$ in magenta for $0 \leq v < +\infty$ in the complex plane in cyan.

$$E_G = E_{G1} + E_{G2} + E_{G3} = E_{G1} \pm i|E_{G2}| + hf.$$

And we can notice that E_G belongs to the set of complex numbers denoted in mathematics by $\mathbb{C}$. Additionally, $|E_{G2}|$ is the module or the norm of the imaginary energy E_{G2} in the metauniverse $\mathbf{G_2}$. The following graph illustrates these facts (**Figure 7**).

4.4 Time intervals and imaginary time

When $v > c$, we get:

$$T' = \frac{\pm i \times T}{\sqrt{\frac{v^2}{c^2} - 1}}$$

If $T' = \frac{+i \times T}{\sqrt{\frac{v^2}{c^2}-1}}$ then this means that when v increases, T' decreases (time contraction).

And if $T' = \frac{-i \times T}{\sqrt{\frac{v^2}{c^2}-1}}$ then this means that when v increases, T' increases (time dilation).

Figure 8.
The flow of time in both the universe $\mathbf{G_1}$ and in a part of the metauniverse $\mathbf{G_2}$.

Concerning the explanation of this is that firstly time goes clockwise in the new four-dimensional continuum $\mathbf{G_2}$ relative to the universe $\mathbf{G_1}$ since it is positive, and secondly it goes counterclockwise relative to the universe $\mathbf{G_1}$ since it is negative. It is to say once more that the imaginary number "i" identifies the new four dimensions that define $\mathbf{G_2}$ (**Figure 8**).

4.5 The real and imaginary lengths

When $v > c$, we will have:

$$L' = \pm i \times L\sqrt{\frac{v^2}{c^2} - 1}$$

If $L' = +i \times L\sqrt{\frac{v^2}{c^2} - 1}$ this means that when v increases so L' increases (Length dilation).

And if $L' = -i \times L\sqrt{\frac{v^2}{c^2} - 1}$ this means that when v increases so L' decreases (Length contraction).

In fact, the minus sign confirms the fact that a length contraction can occur in $\mathbf{G_2}$ when $v > c$ similar to the length contraction in the region where $v < c$ that means in the universe $\mathbf{G_1}$.

4.6 The entropy and the metaentropy

To understand the meaning of negative time in $\mathbf{G_2}$ relative to $\mathbf{G_1}$, then entropy is the best tool. We know that entropy is defined as $dS \geq 0$ in the second principle of thermodynamics. We say that when time grows, then entropy increases. Due to the fact that time is negative as one possible solution in $\mathbf{G_2}$, this implies that we can have $dS \leq 0$. Consequently, and for this case, we say that when time flows, then entropy (or metaentropy) decreases. This means directly the following: The direction of evolution in a part of $\mathbf{G_2}$ is the opposite to that in $\mathbf{G_1}$.

4.7 The transformation of velocities

We have from special relativity: $v = \frac{V' - V}{\left(\frac{VV'}{c^2} - 1\right)} \Leftrightarrow v = \frac{c^2(V' - V)}{(VV' - c^2)}$.

First Case:

This is the case of two bodies in $\mathbf{G_1}$, where their velocities are smaller than c.

$\overrightarrow{G_1}$ $\overrightarrow{G_1}$

We note that: $V = fc$, where $0 \leq f < 1$ and $V' = f'c$, where $0 \leq f' < 1$.
This implies that:

$$v = \frac{c^2(f'c - fc)}{(ff'c^2 - c^2)} = \frac{c^2 \times c \times (f' - f)}{c^2 \times (ff' - 1)} = \frac{c(f' - f)}{ff' - 1}$$

This relation is the one we use in relativistic computations. So, it is not new to us and just as predicted by special relativity.

Second Case:

This is the case of a body in $\mathbf{G_1}$ (where the velocity is $< c$) and a beam of light (where the velocity is c).

$$\overrightarrow{\text{Light}} \qquad \qquad \overrightarrow{G_1}$$

We have now: $V = c$ and $V' = f'c$, where $0 \leq f' < 1$. Then:

$$v = \frac{c^2(f'c - c)}{((c \times f'c) - c^2))} = \frac{c^3 \times (f' - 1)}{c^2 \times (f' - 1)} = c$$

This means that light is the limit velocity in $\mathbf{G_1}$ and is constant in it whatever the velocity of the body in $\mathbf{G_1}$ relative to the beam of light. So just like Albert Einstein's special relativity has predicted.

Third Case:

This is the case of a beam of light relative to another beam of light.

$$\overrightarrow{\text{Light}} \qquad \qquad \overrightarrow{\text{Light}}$$

$$\lim_{V' \to c} v = \lim_{V' \to c} \left[\frac{c^2(V' - V)}{(VV' - c^2)}\right] = \lim_{f' \to 1} \left[\frac{c \times (f' - 1)}{(f' - 1)}\right] = c$$

We know that we have here $V = c$ and $V' = c$. In this case, $v = c$, this is not new to us also. It is the consequence of the relativistic transformation also.

Fourth Case:

This is the case of a beam of light relative to a moving body in $\mathbf{G_2}$, where the velocity is greater than c.

$$\overrightarrow{\text{Light}} \qquad \qquad \overrightarrow{G_2}$$

We have now: $V = c$ and $V' = f'c$, where $1 < f' < +\infty$. Then:

$$v = \frac{c^2(f'c - c)}{((c \times f'c) - c^2))} = \frac{c^3 \times (f' - 1)}{c^2 \times (f' - 1)} = c$$

This means that relative to $\mathbf{G_2}$, light is still the limit velocity and is still constant. In other words, $\mathbf{G_2}$ relative to Light is similar to $\mathbf{G_1}$ relative to Light. Light is the limit velocity in both $\mathbf{G_1}$ and $\mathbf{G_2}$. This fact will be more clarified and more understood in the fifth case.

Fifth Case:

This is the case of a moving body in $\mathbf{G_2}$ relative to another moving body in $\mathbf{G_2}$.

$$\overrightarrow{G_2} \qquad \overrightarrow{G_2}$$

We have from Metarelativity: $m = \frac{\pm i \times m_0}{\sqrt{\frac{v^2}{c^2}-1}} \Leftrightarrow m_{G2} = \frac{m_{0,G2}}{\sqrt{\frac{v^2}{c^2}-1}}$.where $m_{0,G2}$ is the starting or the smallest mass in $\mathbf{G_2}$.

$$\text{So, if } m_{G2} = m_{0,G2} \Leftrightarrow m_{0,G2} = \frac{m_{0,G2}}{\sqrt{\frac{V^2}{c^2}-1}} \Leftrightarrow 1 = \frac{1}{\sqrt{\frac{V^2}{c^2}-1}}$$

$$\Leftrightarrow \sqrt{\frac{V^2}{c^2}-1} = 1 \Leftrightarrow \frac{V^2}{c^2} - 1 = 1^2 = 1 \Leftrightarrow \frac{V^2}{c^2} = 2 \Leftrightarrow V^2 = 2c^2$$

$$\Leftrightarrow V = c\sqrt{2}$$

That means that the starting mass which is $m_{0,G2}$ in the metauniverse $\mathbf{G_2}$ occurs when $V = c\sqrt{2}$.

Assume that $V = c\sqrt{2}$ (the smallest velocity in $\mathbf{G_2}$) and that $V' = \beta c\sqrt{2}$ that is any velocity greater or equal to the starting velocity, in other words: $\beta \geq 1$. This implies that:

$$v = \frac{c^2\left(\beta c\sqrt{2} - c\sqrt{2}\right)}{(2\beta c^2 - c^2)} = \frac{c\sqrt{2}(\beta - 1)}{(2\beta - 1)}$$

If $\beta = 1$ then $v = 0$.

If $\beta \to +\infty$ then $v \to \frac{c\sqrt{2}}{2} = 0.7071c < c$.

This is similar to the relativistic transformations since we have:
$0 \leq v \leq \frac{c\sqrt{2}}{2} = 0.7071c < c$. As if we are working in $\mathbf{G_1}$ exactly. This means that the universe $\mathbf{G_2}$ relatively to itself behaves like the universe $\mathbf{G_1}$ relative to itself since the velocity of $\mathbf{G_2}$ relative to $\mathbf{G_2}$ is smaller than c just like the velocity of $\mathbf{G_1}$ relative to $\mathbf{G_1}$. This fact can be also explained as follows: $\mathbf{G_2}$ is as real as $\mathbf{G_1}$ relative to itself but at a different level of experience and in higher dimensions. Accordingly, we can say that $\mathbf{G_2}$ relative to itself is a "real" universe but relative to $\mathbf{G_1}$ is an "imaginary" universe as it will be shown in the sixth and seventh cases.

Sixth Case:

This is the case of $\mathbf{G_2}$ relative to $\mathbf{G_1}$.

$$\overrightarrow{G_1} \qquad \overrightarrow{G_2}$$

We note that: $V = fc$, where $0 \leq f < 1$ and $V' = f'c$, where $1 < f' < +\infty$.

This implies that:

$$v = \frac{c^2(f'c - fc)}{(ff'c^2 - c^2)} = \frac{c^2 \times c \times (f' - f)}{c^2 \times (ff' - 1)} = \frac{c(f' - f)}{ff' - 1}$$

So, if $f = 0 \Rightarrow v = \frac{c(f'-0)}{(0 \times f')-1} = \frac{f'c}{-1} = -f'c \Rightarrow |v| = f'c > c$ since $1 < f' < +\infty$.

And if $f \to 1 \Rightarrow v \to \frac{c(f'-1)}{(1 \times f')-1} = \frac{c(f'-1)}{f'-1} = c$.

We have here $|v| > c$. This implies that the metarelativistic transformations are needed here and for the first time.

Seventh Case:

This is the case of $\mathbf{G_1}$ relative to $\mathbf{G_2}$.

$$\overrightarrow{G_2} \qquad\qquad \overrightarrow{G_1}$$

We note that: $V = fc$, where $1 < f < +\infty$ and $V' = f'c$, where $0 \le f' < 1$.

This implies that:

$$v = \frac{c^2(f'c - fc)}{(ff'c^2 - c^2)} = \frac{c^2 \times c \times (f' - f)}{c^2 \times (ff' - 1)} = \frac{c(f' - f)}{ff' - 1}$$

So, if $f' = 0 \Rightarrow v = \frac{c(0-f)}{(f\times 0)-1} = \frac{-fc}{-1} = fc > c$ since $1 < f < +\infty$.

And if $f' \to 1 \Rightarrow v \to \frac{c(1-f)}{(f\times 1)-1} = \frac{(1-f)c}{f-1} = -c \Rightarrow |v| \to c$.

We have here $|v| > c$. This implies that the metarelativistic transformations are needed here also.

4.8 The new principle of Metarelativity

Now, if we want to elaborate on the new principle of Metarelativity, it will be:

"Inertial observers must correlate their observations by means of relativistic Lorentz transformations if the velocity is smaller than c and by means of the metarelativistic transformations if the velocity is greater than c, and all physical quantities must transform from one inertial system to another in such a way that the expression of the physical laws is the same for all inertial observers. The subluminal universe is denoted by $\mathbf{G_1}$, the superluminal universe is denoted by $\mathbf{G_2}$, and the luminal universe of frequencies is denoted by $\mathbf{G_3}$. The sum of $\mathbf{G_1}$, of the electromagnetic waves **EW**, and of $\mathbf{G_2}$ is denoted by:

$$\mathbf{G} = \mathbf{G_1} + \mathbf{G_3} + \mathbf{G_2} = \mathbf{G_1} + \textbf{\textit{Light}} + \mathbf{G_2}$$

and all electromagnetic waves (that include light) are at constant velocity in both $\mathbf{G_1}$ and $\mathbf{G_2}$."

As it was shown that the new theory does not destroy Einstein's theory of relativity that we know at all but on the contrary, it proves its veracity and then expands it to the set of complex masses, time, lengths, and energies, which is the eight-dimensional complex hyperspace $\mathbb{C}$ or equivalently in the total universe **G**.

5. The four models of *MCPP*: The road to the final and most general model of *MCPP*

In this work and in the following sections in chapters 1 and 2 we will consider three simplified models of *MCPP* then present at the end the final and most general model [1–24, 30–43]. Each model is an enhanced and wider model than the previous one. In all four models, we will consider the velocities of the bodies moving in **G** to be random variables that follow certain probability distributions (*PDFs*) and certain corresponding cumulative probability distribution functions (*CDFs*) in both the subluminal universe $\mathbf{G_1}$ and the superluminal metauniverse $\mathbf{G_2}$. I have followed this methodology in order to develop gradually and systematically the *MCPP* paradigm and in order to reach the final and most general model of *MCPP* that can be adopted in any possible and imaginable situation. Even the deterministic case, which is a special case of the general random *CPP,* was also presented and considered in order to show that *MCPP* is valid everywhere either in the deterministic or in the random case. Consequently, and in each model, we evaluate the associated real, imaginary, and complex probabilities as well as all the related *MCPP* parameters in the probabilities sets $\mathcal{R}$, $\mathcal{M}$, and $\mathcal{C}$; hence, in the universes $\mathbf{G_1}$, $\mathbf{G_2}$, $\mathbf{G_3}$, and $\mathbf{G}$. Thus, we connect successfully *CPP* with Metarelativity to unify both theories in a general and a unified paradigm that we called *MCPP*.

In the first simplified model, the body velocities PDF_1 in $\mathbf{G_1}$ and PDF_2 in $\mathbf{G_2}$ are taken to be both Gaussian and normal, in addition, the velocity in the metauniverse $\mathbf{G_2}$ varies here between c (light velocity) and $2c$. In this reduced model, we restricted our study to $\mathbf{G} = \mathbf{G_1} + \mathbf{G_2}$.

In the second simplified and more general model, the body velocities PDF_1 in $\mathbf{G_1}$ and PDF_2 in $\mathbf{G_2}$ are taken also to be both Gaussian and normal, in addition, the velocity in $\mathbf{G_2}$ varies between c and nc, where n is an arbitrary and predetermined number having: $\forall n, n \in \mathbb{R}^+ : n > 1 \Leftrightarrow n \in (1, +\infty)$. This was done since the velocities in $\mathbf{G_2}$ according to Metarelativity vary between c and infinity. Hence, the second model is an improved version of the first one. In this model, we restricted our study also to $\mathbf{G} = \mathbf{G_1} + \mathbf{G_2}$.

In the third simplified and wider model, the body velocities PDF_1 in $\mathbf{G_1}$ and PDF_2 in $\mathbf{G_2}$ are taken to follow any possible probability distribution whether discrete or continuous and they do not have to be similar at all like in the previous two models, this in order to be realistic. In addition, the velocity in $\mathbf{G_2}$ varies between c and nc also. Thus, the third model is an enhanced model. In this model, we have considered also only the universe $\mathbf{G} = \mathbf{G_1} + \mathbf{G_2}$.

The final and most general model is the sought model of *MCPP*. It is the goal of all the calculations made and of the methodology adopted. Here, the body velocities PDF_1 in $\mathbf{G_1}$ and PDF_2 in $\mathbf{G_2}$ are taken to follow any possible probability distribution whether discrete or continuous and they do not have to be similar at all in order to be totally realistic. The velocity in $\mathbf{G_2}$ varies between c and nc also. Additionally, we have included the contributions of the luminal universe of electromagnetic waves, which is $\mathbf{G_3}$, where the velocity of the EW is c and the corresponding frequency follows any possible probability distribution PDF_3 and CDF_3, respectively. Therefore, we have considered here the most general case which is the total universe $\mathbf{G} = \mathbf{G_1} + \mathbf{G_2} + \mathbf{G_3}$. This final model links definitively and in the most general way *CPP* with Metarelativity into the unified paradigm of *MCPP*.

Furthermore, in all four models in the two chapters, we have defined, calculated, simulated, illustrated, and drawn all the probabilities and all the *MCPP* parameters in $\mathcal{R}$, $\mathcal{M}$, and $\mathcal{C} = \mathbf{G}$.

6. The Metarelativistic complex probability paradigm (*MCPP*): A first model

6.1 The real and imaginary probabilities

Let v_1 be the velocity of a body in R_1 with $0 \leq v_1 < c$ and let it be a random variable that follows the normal distribution: $N(\overline{v}_1 = c/2, \sigma_{v1} = c/6)$ where $\overline{v}_1$ is the mean or the expectation of this symmetric normal probability distribution of v_1 or $PDF_1(v_1)$ and σ_{v1} is its corresponding standard deviation.

And let v_2 be the velocity of a body in R_2 with $c < v_2 \leq 2c$ and let it be a random variable that follows the normal distribution: $N(\overline{v}_2 = 3c/2, \sigma_{v2} = c/6)$ where $\overline{v}_2$ is the mean or the expectation of this symmetric normal probability distribution of v_2 or $PDF_2(v_2)$ and σ_{v2} is its corresponding standard deviation.

Then, $P_{R1} = P_{rob}(0 \leq V \leq v_1) = CDF_1(0 \leq V \leq v_1) = \int_0^{v1} PDF_1(v)dv = \int_0^{v1} N(\overline{v} = c/2, \sigma_v = c/6)dv.$

If $v_1 < 0 \Rightarrow P_{R1} = P_{rob}(V < 0) = CDF_1(V < 0) = 0.$

If $v_1 = 0 \Rightarrow P_{R1} = P_{rob}(V \leq 0) = CDF_1(V \leq 0) = \int_0^0 PDF_1(v)dv = 0.$

If $v_1 = \overline{v}_1 = c/2 \Rightarrow P_{R1} = P_{rob}(0 \leq V \leq c/2) = CDF_1(0 \leq V \leq c/2) = \int_0^{c/2} PDF_1(v)dv = 0.5.$

If $v_1 \to c^- \Rightarrow P_{R1} \to P_{rob}(0 \leq V < c) = CDF_1(0 \leq V < c) = \int_0^c PDF_1(v)dv = \int_0^c N(\overline{v} = c/2, \sigma_v = c/6)dv = 1.$

If $v_1 > c \Rightarrow P_{R1} = P_{rob}(V > c) = CDF_1(V > c) = \int_0^{v1} PDF_1(v)dv = \left\{\int_0^c PDF_1(v)dv + \int_c^{v1} PDF_1(v)dv\right\} = (1 + 0) = 1.$

And $P_{R2} = P_{rob}(c < V \leq v_2) = CDF_2(c < V \leq v_2) = \int_c^{v2} PDF_2(v)dv = \int_c^{v2} N(\overline{v} = 3c/2, \sigma_v = c/6)dv.$

If $v_2 < c \Rightarrow P_{R2} = P_{rob}(V < c) = CDF_2(V < c) = 0.$

If $v_2 \to c^+ \Rightarrow P_{R2} \to \int_c^c PDF_2(v)dv = \int_c^c N(\overline{v} = 3c/2, \sigma_v = c/6)dv = 0.$

If $v_2 = \overline{v}_2 = 3c/2 \Rightarrow P_{R2} = P_{rob}(c < V \leq 3c/2) = CDF_2(c < V \leq 3c/2) = \int_c^{3c/2} PDF_2(v)dv = 0.5.$

If $v_2 = 2c \Rightarrow P_{R2} = P_{rob}(c < V \leq 2c) = CDF_2(c < V \leq 2c) = \int_c^{2c} PDF_2(v)dv = \int_c^{2c} N(\overline{v} = 3c/2, \sigma_v = c/6)dv = 1.$

If $v_2 > 2c$

$$\Rightarrow P_{R2} = P_{rob}(V > 2c) = CDF_2(V > 2c)$$
$$= \int_c^{v2} PDF_2(v)dv = \left\{\int_c^{2c} PDF_2(v)dv + \int_{2c}^{v2} PDF_2(v)dv\right\} = (1 + 0) = 1$$

Moreover,

$$P_{M1} = i(1 - P_{R1}) = i[1 - P_{rob}(0 \leq V \leq v_1)] = i[1 - CDF_1(0 \leq V \leq v_1)] = iCDF_1(v_1 < V < c)$$
$$= i\left[1 - \int_0^{v1} PDF_1(v)dv\right] = i\int_{v1}^{c} PDF_1(v)dv = i\int_{v1}^{c} N(\overline{v} = c/2, \sigma_v = c/6)dv$$

If $v_1 < 0 \Rightarrow$

$$P_{M1} = i\int_{v1}^{c} PDF_1(v)dv = i\left\{\int_{v1}^{0} PDF_1(v)dv + \int_0^c PDF_1(v)dv\right\} = i(0 + 1) = i \Rightarrow P_{M1}/i = 1.$$

If $v_1 = 0$

$\Rightarrow P_{M1} = i[1 - P_{rob}(V \leq 0)] = i[1 - CDF_1(V \leq 0)] = i(1 - 0) = i \Rightarrow P_{M1}/i = 1.$

If $v_1 = \bar{v}_1 = c/2$

$$\Rightarrow P_{M1} = i[1 - P_{rob}(0 \le V \le c/2)] = i[1 - CDF_1(0 \le V \le c/2)]$$
$$= i\left[1 - \int_0^{c/2} PDF_1(v)dv\right] = i\int_{c/2}^{c} PDF_1(v)dv = i(1-0.5) = 0.5i \Rightarrow P_{M1}/i = 0.5$$

If $v_1 \to c^-$
$$\Rightarrow P_{M1} \to i[1 - P_{rob}(0 \le V < c)] = i[1 - CDF_1(0 \le V < c)] = i\left[1 - \int_0^c PDF_1(v)dv\right] = i(1-1) = 0$$
$\Rightarrow P_{M1}/i \to 0$
If $v_1 > c \Rightarrow P_{M1} = 0 \Rightarrow P_{M1}/i = 0$.
And

$$P_{M2} = i(1 - P_{R2}) = i[1 - P_{rob}(c < V \le v_2)] = i[1 - CDF_2(c < V \le v_2)] = iCDF_2(v_2 < V \le 2c)$$
$$= i\left[1 - \int_c^{v2} PDF_2(v)dv\right] = i\int_{v2}^{2c} PDF_2(v)dv = i\int_{v2}^{2c} N(\bar{v} = 3c/2, \sigma_v = c/6)dv$$

If $v_2 < c \Rightarrow P_{M2} = i\int_{v2}^{2c} PDF_2(v)dv = i\left\{\int_{v2}^{c} PDF_2(v)dv + \int_c^{2c} PDF_2(v)dv\right\} = i(0+1) = i$
$\Rightarrow P_{M2}/i = 1$.
If
$v_2 \to c^+ \Rightarrow P_{M2} \to i[1 - P_{rob}(V \le v_2)] = iP_{rob}(c < V \le 2c) = i \times 1 = i \Rightarrow P_{M2}/i \to 1$.
If $v_2 = \bar{v}_2 = 3c/2$

$$\Rightarrow P_{M2} = i[1 - P_{rob}(c < V \le 3c/2)] = i[1 - CDF_2(c < V \le 3c/2)]$$
$$= i\left[1 - \int_c^{3c/2} PDF_2(v)dv\right] = i\int_{3c/2}^{2c} PDF_2(v)dv = i(1-0.5) = 0.5i \Rightarrow P_{M2}/i = 0.5$$

If $v_2 = 2c$
$$\Rightarrow P_{M2} = i[1 - P_{rob}(c < V \le 2c)] = i[1 - CDF_2(c < V \le 2c)] = i\left[1 - \int_c^{2c} PDF_2(v)dv\right] = i(1-1) = 0$$
$\Rightarrow P_{M2}/i = 0$
If $v_2 > 2c \Rightarrow P_{M2} = 0 \Rightarrow P_{M2}/i = 0$.
We have $\mathcal{R} = R_1(0 \le v < c) + R_2(c < v \le 2c)$.
Now, let $P_R = \frac{P_{R1}+P_{R2}}{2}$ and it is equal to half of the sum of the cumulative probability that $0 \le V \le v_1$ in R_1 and the cumulative probability that $c < V \le v_2$ in R_2.

$$\Rightarrow P_R = \frac{CDF_1(0 \le V \le v_1) + CDF_2(c < V \le v_2)}{2}$$
$$= \frac{1}{2}\left\{\int_0^{v1} PDF_1(v)dv + \int_c^{v2} PDF_2(v)dv\right\}$$
$$= \frac{1}{2}\left\{\int_0^{v1} N(\bar{v} = c/2, \sigma_v = c/6)dv + \int_c^{v2} N(\bar{v} = 3c/2, \sigma_v = c/6)dv\right\}$$

We have in $\mathbf{G} = \mathcal{C} = \mathcal{R} + \mathcal{M} = \mathbf{G_1} + \mathbf{G_2}$: $0 \le v \le 2c$ with $v \neq c$.
So, if $0 \le v < c \Rightarrow P_{R1} = P_{rob}(0 \le V \le v) = CDF_1(0 \le V \le v)$.
And $P_{R2} = P_{rob}(V < c) = CDF_2(V < c) = 0$

$$\Rightarrow P_R = \frac{CDF_1(0 \le V \le v) + 0}{2} = \frac{CDF_1(0 \le V \le v)}{2} = \frac{P_{R1}}{2}$$

Therefore, we say here that we are working in the real probability universe $\mathcal{R} = R_1$ alone.
And if $c < v \le 2c \Rightarrow P_{R1} = P_{rob}(V > c) = CDF_1(V > c) = 1$.
And $P_{R2} = P_{rob}(c < V \le v) = CDF_2(c < V \le v)$

$$\Rightarrow P_R = \frac{1 + CDF_2(c < V \le v)}{2} = \frac{1 + P_{R2}}{2}$$

Therefore, we say here that we are working in the real probability universe $\mathcal{R} = R_2$ alone.
And, if $0 \le v \le 2c$ with $v \neq c \Rightarrow P_{R1} = P_{rob}(0 \le V \le v) = CDF_1(0 \le V \le v)$.
And $P_{R2} = P_{rob}(c < V \le v) = CDF_2(c < V \le v)$

$$\Rightarrow P_R = \frac{CDF_1(0 \le V \le v) + CDF_2(c < V \le v)}{2} = \frac{P_{R1} + P_{R2}}{2}$$

Therefore, we say here that we are working in the real probability universe $\mathcal{R} = R_1 + R_2$.
And consequently,
if $v < 0 \Rightarrow P_R = \frac{CDF_1(V < 0)}{2} = \frac{0}{2} = 0$.
if $v = c/2 \Rightarrow P_R = \frac{CDF_1(0 \le V \le c/2) + CDF_2(V < c)}{2} = \frac{0.5 + 0}{2} = 0.25$.
if $v \to c^- \Rightarrow P_R \to \frac{CDF_1(0 \le V < c) + CDF_2(V < c)}{2} = \frac{1+0}{2} = 0.5$.
if $v = 3c/2 \Rightarrow P_R = \frac{CDF_1(0 \le V < c) + CDF_2(c < V \le 3c/2)}{2} = \frac{1 + 0.5}{2} = 0.75$.
if $v = 2c \Rightarrow P_R = \frac{CDF_1(0 \le V < c) + CDF_2(c < V \le 2c)}{2} = \frac{1+1}{2} = 1$.
We have $\mathcal{M} = M_1(0 \le v < c) + M_2(c < v \le 2c)$.
Now, let $P_M = \frac{P_{M1} + P_{M2}}{2}$ and it is equal to half of the sum of the complement of the cumulative probability that $0 \le V \le v_1$ in M_1 and the complement of the cumulative probability that $c < V \le v_2$ in M_2.

$$\begin{aligned}
\Rightarrow P_M &= \frac{i(1 - P_{R1}) + i(1 - P_{R2})}{2} \\
&= \frac{2i - i(P_{R1} + P_{R2})}{2} = i - \frac{i(P_{R1} + P_{R2})}{2} = i\left[1 - \frac{(P_{R1} + P_{R2})}{2}\right] = i(1 - P_R) \\
\Rightarrow P_M &= \frac{i[1 - CDF_1(0 \le V \le v_1)] + i[1 - CDF_2(c < V \le v_2)]}{2} \\
&= \frac{i}{2}\left\{\int_0^{v1} [1 - PDF_1(v)]dv + \int_c^{v2} [1 - PDF_2(v)]dv\right\} \\
&= \frac{i}{2}\left\{\int_0^{v1} [1 - N(\overline{v} = c/2, \sigma_v = c/6)]dv + \int_c^{v2} [1 - N(\overline{v} = 3c/2, \sigma_v = c/6)]dv\right\} \\
&= \frac{i}{2}\left\{\int_{v1}^{c} N(\overline{v} = c/2, \sigma_v = c/6)dv + \int_{v2}^{2c} N(\overline{v} = 3c/2, \sigma_v = c/6)dv\right\}
\end{aligned}$$

We have in $\mathbf{G} = \mathcal{C} = \mathcal{R} + \mathcal{M} = \mathbf{G_1} + \mathbf{G_2}$: $0 \le v \le 2c$ with $v \ne c$.
So, if $0 \le v < c \Rightarrow P_{M1} = i[1 - P_{rob}(0 \le V \le v)] = i[1 - CDF_1(0 \le V \le v)]$.
And $P_{M2} = i[1 - P_{rob}(V < c)] = i[1 - CDF_2(V < c)] = i(1 - 0) = i$

$$\Rightarrow P_M = \frac{i[1 - CDF_1(0 \le V \le v)] + i}{2} = \frac{i + P_{M1}}{2} = i\left[1 - \frac{P_{R1}}{2}\right]$$

Therefore, we say here that we are working in the imaginary probability universe $\mathcal{M} = M_1$ alone.
And if $c < v \le 2c \Rightarrow P_{M1} = i[1 - P_{rob}(V > c)] = i[1 - CDF_1(V > c)] = i(1 - 1) = 0$.
And $P_{M2} = i[1 - P_{rob}(c < V \le v)] = i[1 - CDF_2(c < V \le v)]$

$$\Rightarrow P_M = \frac{0 + i[1 - CDF_2(c < V \le v)]}{2} = \frac{P_{M2}}{2} = i\left[\frac{1 - P_{R2}}{2}\right]$$

Therefore, we say here that we are working in the imaginary probability universe $\mathcal{M} = M_2$ alone.
And, if
$0 \le v \le 2c$ with $v \ne c \Rightarrow P_{M1} = i[1 - P_{rob}(0 \le V \le v)] = i[1 - CDF_1(0 \le V \le v)]$.
And $P_{M2} = i[1 - P_{rob}(c < V \le v)] = i[1 - CDF_2(c < V \le v)]$
$\Rightarrow P_M = \frac{i[1-CDF_1(0 \le V < c)] + i[1-CDF_2(c < V \le v)]}{2} = \frac{P_{M1}+P_{M2}}{2} = i\left[1 - \frac{P_{R1}+P_{R2}}{2}\right] = i[1 - P_R]$.
Therefore, we say here that we are working in the imaginary probability universe $\mathcal{M} = M_1 + M_2$.
And consequently, if $v < 0 \Rightarrow P_M = i\left[1 - \frac{CDF_1(V<0)}{2}\right] = i\left[1 - \frac{0}{2}\right] = i \Rightarrow P_M/i = 1$.
if $v = c/2 \Rightarrow P_M = i\left[1 - \frac{CDF_1(0 \le V \le c/2) + CDF_2(V < c)}{2}\right] = i\left[1 - \frac{0.5+0}{2}\right] = 0.75i$

$$\Rightarrow P_M/i = 0.75$$

if $v \to c^- \Rightarrow P_M \to i\left[1 - \frac{CDF_1(0 \le V < c) + CDF_2(V < c)}{2}\right] = i\left[1 - \frac{1+0}{2}\right] = 0.5i \Rightarrow P_M/i \to 0.5$.
if $v = 3c/2 \Rightarrow P_M = i\left[1 - \frac{CDF_1(0 \le V < c) + CDF_2(c < V \le 3c/2)}{2}\right] = i\left[1 - \frac{1+0.5}{2}\right] = 0.25i$

$$\Rightarrow P_M/i = 0.25$$

if $v = 2c \Rightarrow P_M = i\left[1 - \frac{CDF_1(0 \le V < c) + CDF_2(c < V \le 2c)}{2}\right] = i\left[1 - \frac{1+1}{2}\right] = i(1-1) = 0$

$$\Rightarrow P_M/i = 0$$

In addition, since $v = c$ is an axis of symmetry then we can deduce from calculus that:

$$P_{R1}[0 \le v_1 < c] = P_{M2}[c < (v_2 = 2c - v_1) \le 2c]/i = 1 - P_{R2}[c < (v_2 = 2c - v_1) \le 2c]$$

Check that: $P_{R1}[0 \le (v_1 = 0) < c] = P_{M2}[c < (v_2 = 2c - v_1 = 2c - 0 = 2c) \le 2c]/i = 0$

$$P_{R1}[0 \le (v_1 = c/2) < c] = P_{M2}[c < (v_2 = 2c - v_1 = 2c - c/2 = 3c/2) \le 2c]/i = 0.5$$
$$P_{R1}[0 \le (v_1 \to c) < c] = P_{M2}[c < (v_2 = 2c - v_1 \to 2c - c = c) \le 2c]/i = 1$$

And,

$$P_{R2}[c < v_2 \leq 2c] = P_{M1}[0 \leq (v_1 = 2c - v_2) < c]/i = 1 - P_{R1}[0 \leq (v_1 = 2c - v_2) < c]$$

Check that: $P_{R2}[c < (v_2 = 2c) \leq 2c] = P_{M1}[0 \leq (v_1 = 2c - v_2 = 2c - 2c = 0) < c]/i = 1$

$$P_{R2}[c < (v_2 = 3c/2) \leq 2c] = P_{M1}[0 \leq (v_1 = 2c - v_2 = 2c - 3c/2 = c/2) < c]/i = 0.5$$
$$P_{R2}[c < (v_2 \to c) \leq 2c] = P_{M1}[0 \leq (v_1 = 2c - v_2 \to 2c - c = c) < c]/i = 0$$

Therefore, for any value of $0 \leq v \leq 2c$ with $v \neq c$, we can write without any confusion that:

$$P_{M1} = i(1 - P_{R1}) \text{ and } P_{R1} = 1 - P_{M1}/i$$

hence, M_1 is the imaginary complementary probability universe to the real probability universe R_1.

$$\text{And } P_{M2} = i(1 - P_{R2}) \text{ and } P_{R2} = 1 - P_{M2}/i$$

hence, M_2 is the imaginary complementary probability universe to the real probability universe R_2.

Additionally, in condition and in the case where $v_2 = 2c - v_1$, we have:

$$P_{R1}[0 \leq v_1 < c] + P_{R2}[c < (v_2 = 2c - v_1) \leq 2c] = 1$$

hence, R_2 is the real complementary probability universe to the real probability universe R_1.

$$\text{And, } P_{M1}[0 \leq v_1 < c] + P_{M2}[c < (v_2 = 2c - v_1) \leq 2c] = i$$

hence, M_2 is the imaginary complementary probability universe to the imaginary probability universe M_1.

Moreover, in all cases and for any value of $v : 0 \leq v \leq 2c$ with $v \neq c$, we have:

$P_R = \frac{P_{R1}+P_{R2}}{2}$ where $\mathcal{R} = R_1 + R_2$.

And $P_M = \frac{P_{M1}+P_{M2}}{2}$ where $\mathcal{M} = M_1 + M_2$.

Check that:

$$P_M = \frac{i(1 - P_{R1}) + i(1 - P_{R2})}{2}$$
$$= \frac{2i - i(P_{R1} + P_{R2})}{2} = i - \frac{i(P_{R1} + P_{R2})}{2} = i\left[1 - \frac{(P_{R1} + P_{R2})}{2}\right] = i(1 - P_R)$$

Hence, $\mathcal{M}$ is the imaginary complementary probability universe to the real probability universe $\mathcal{R}$.

Moreover, we have in $\mathbf{G} = \mathcal{C} = \mathcal{R} + \mathcal{M}$, where $0 \leq v \leq 2c$ with $v \neq c$:

$\mathcal{C} = (R_1 + R_2) + (M_1 + M_2) = (R_1 + M_1) + (R_2 + M_2) = \mathcal{C}_1 + \mathcal{C}_2$.

In fact, in $\mathcal{C}_1$ we have: $Pc_1 = P_{R1} + P_{M1}/i = P_{R1} + (1 - P_{R1}) = 1$.

And, in $\mathcal{C}_2$ we have: $Pc_2 = P_{R2} + P_{M2}/i = P_{R2} + (1 - P_{R2}) = 1$.

And, in $\mathcal{C}$ we have:

$$Pc = P_R + P_M/i = \frac{P_{R1} + P_{R2}}{2} + \left[\frac{P_{M1} + P_{M2}}{2}\right]/i$$
$$= \frac{P_{R1} + P_{R2}}{2} + \left[\frac{i(1 - P_{R1}) + i(1 - P_{R2})}{2}\right]/i$$
$$= \frac{P_{R1} + P_{R2}}{2} + \frac{(1 - P_{R1}) + (1 - P_{R2})}{2} = \frac{P_{R1} + P_{R2}}{2} + 1 - \frac{P_{R1} + P_{R2}}{2}$$
$$= 1$$

We can write also:

$$
\begin{aligned}
Pc &= P_R + P_M/i \\
&= \frac{P_{R1}+P_{R2}}{2} + \left[\frac{P_{M1}+P_{M2}}{2}\right]/i = \frac{P_{R1}+P_{M1}/i}{2} + \frac{P_{R2}+P_{M2}/i}{2} = \frac{Pc_1}{2} + \frac{Pc_2}{2} \\
&= \frac{Pc_1+Pc_2}{2} = \frac{1+1}{2} = 1
\end{aligned}
$$

Consequently: $Pc = Pc_1 = Pc_2 = 1$, in accordance with *CPP* axioms.

Furthermore, we have:

$\mathbf{G} = \mathbf{G_1}\ (0 \le v < c) + \mathbf{G_2}\ (c < v \le 2c)$, which means that the total universe **G** is the sum of the real subluminal universe $\mathbf{G_1}$ and the imaginary superluminal universe or metauniverse $\mathbf{G_2}$.

Additionally, the real subluminal universe $\mathbf{G_1}$ corresponds to the complex probability universe $\mathcal{C}_1$, which is also subluminal, hence: $\mathbf{G_1} = \mathcal{C}_1 = R_1 + M_1$ with $(0 \le v < c)$. And the imaginary superluminal universe $\mathbf{G_2}$ or metauniverse corresponds to the complex probability universe $\mathcal{C}_2$, which is also superluminal; hence, $\mathbf{G_2} = \mathcal{C}_2 = R_2 + M_2$ with $(c < v \le 2c)$.

Therefore,

$$
\begin{aligned}
P_{G1} &= Pc_1 = P_{R1} + P_{M1}/i = P_{R1} + (1 - P_{R1}) = 1 \text{ and} \\
P_{G2} &= Pc_2 = P_{R2} + P_{M2}/i = P_{R2} + (1 - P_{R2}) = 1.
\end{aligned}
$$

Consequently, the complex total universe $\mathbf{G} = \mathbf{G_1}\ (0 \le v < c) + \mathbf{G_2}\ (c < v \le 2c)$, which is the sum of the universe and the metauniverse corresponds to the complex probability universe $\mathcal{C}$ having:

$\mathbf{G} = \mathcal{C} = \mathcal{R} + \mathcal{M} = (R_1 + R_2) + (M_1 + M_2) = (R_1 + M_1) + (R_2 + M_2)$
$= \mathcal{C}_1\ (0 \le v < c) + \mathcal{C}_2\ (c < v \le 2c) = \mathbf{G_1}\ (0 \le v < c) + \mathbf{G_2}\ (c < v \le 2c)$ with $0 \le v \le 2c$ and $v \ne c$,

Hence,

$$
P_G = Pc = \frac{P_{G1}+P_{G2}}{2} = \frac{Pc_1+Pc_2}{2} = \frac{1+1}{2} = 1
$$

Hence, $Pc = 1$, in accordance with *CPP* axioms.

Thus, we can conclude that, by adding the complementary imaginary probabilities universes M_1, M_2 and $\mathcal{M}$ to the real probabilities universes R_1, R_2 and $\mathcal{R}$ then all random phenomena in the complex probabilities' universes $\mathcal{C}_1$, $\mathcal{C}_2$, and $\mathcal{C}$, and hence in the subluminal universe $\mathbf{G_1}$, in the superluminal universe $\mathbf{G_2}$, and in the total and complex universe **G**, become absolutely and perfectly deterministic with probabilities:

$$
Pc = Pc_1 = Pc_2 = 1 \text{ and } P_G = P_{G1} = P_{G2} = 1.
$$

6.2 The *MCPP* parameters of the first model

The real probabilities in $\mathcal{R} = R_1 + R_2$:

$$
P_R = \frac{P_{R1}+P_{R2}}{2}
$$

The imaginary complementary probabilities in $\mathcal{M} = M_1 + M_2$:

$$P_M = \frac{P_{M1} + P_{M2}}{2} = i(1 - P_R) = i\left[1 - \frac{(P_{R1} + P_{R2})}{2}\right]$$

The real complementary probabilities in $\mathcal{R} = R_1 + R_2$:

$$P_M/i = \left[\frac{P_{M1} + P_{M2}}{2}\right]/i = \frac{P_{M1}/i + P_{M2}/i}{2}$$
$$= i(1 - P_R)/i = 1 - \frac{(P_{R1} + P_{R2})}{2}$$

The Complex Random Vectors in $\mathbf{G} = \mathcal{C} = \mathcal{R} + \mathcal{M} = \mathbf{G_1} + \mathbf{G_2}$.
We have: In $\mathbf{G_1}$: $Z_1 = P_{R1} + P_{M1}$ and in $\mathbf{G_2}$: $Z_2 = P_{R2} + P_{M2}$.
Then, in $\mathbf{G}$: $Z = P_R + P_M = \frac{P_{R1}+P_{R2}}{2} + \frac{P_{M1}+P_{M2}}{2} = \frac{P_{R1}+P_{M1}}{2} + \frac{P_{R2}+P_{M2}}{2} = \frac{Z_1+Z_2}{2}$.
The degrees of our knowledge:
We have: In $\mathbf{G_1}$: $DOK_1 = |Z_1|^2 = |P_{R1} + P_{M1}|^2 = P_{R1}^2 + [P_{M1}/i]^2 = P_{R1}^2 + [1 - P_{R1}]^2$.
In $\mathbf{G_2}$: $DOK_2 = |Z_2|^2 = |P_{R2} + P_{M2}|^2 = P_{R2}^2 + [P_{M2}/i]^2 = P_{R2}^2 + [1 - P_{R2}]^2$.
Then, in $\mathbf{G}$:

$$DOK = |Z|^2 = |P_R + P_M|^2 = P_R^2 + [P_M/i]^2 = P_R^2 + [1 - P_R]^2$$
$$= \left[\frac{P_{R1} + P_{R2}}{2}\right]^2 + \left[1 - \left(\frac{P_{R1} + P_{R2}}{2}\right)\right]^2$$

The chaotic factors:
We have: In $\mathbf{G_1}$:
$Chf_1 = 2iP_{R1}P_{M1} = 2iP_{R1}i(1 - P_{R1}) = 2i^2P_{R1}(1 - P_{R1}) = -2P_{R1}(1 - P_{R1})$.
In $\mathbf{G_2}$: $Chf_2 = 2iP_{R2}P_{M2} = 2iP_{R2}i(1 - P_{R2}) = 2i^2P_{R2}(1 - P_{R2}) = -2P_{R2}(1 - P_{R2})$.
Then, in $\mathbf{G}$:

$$Chf = 2iP_RP_M = 2iP_Ri(1 - P_R) = 2i^2P_R(1 - P_R) = -2P_R(1 - P_R)$$
$$= -2\left[\frac{P_{R1} + P_{R2}}{2}\right]\left[1 - \left(\frac{P_{R1} + P_{R2}}{2}\right)\right]$$

The magnitudes of the chaotic factors:
We have: In $\mathbf{G_1}$:
$MChf_1 = |Chf_1| = -2iP_{R1}P_{M1} = -2iP_{R1}i(1 - P_{R1}) = -2i^2P_{R1}(1 - P_{R1}) = 2P_{R1}(1 - P_{R1})$.
In $\mathbf{G_2}$: $MChf_2 = |Chf_2| = -2iP_{R2}P_{M2} = -2iP_{R2}i(1 - P_{R2}) = -2i^2P_{R2}(1 - P_{R2}) = 2P_{R2}(1 - P_{R2})$.
Then, in $\mathbf{G}$:

$$MChf = |Chf| = -2iP_RP_M = -2iP_Ri(1 - P_R) = -2i^2P_R(1 - P_R) = 2P_R(1 - P_R)$$
$$= 2\left[\frac{P_{R1} + P_{R2}}{2}\right]\left[1 - \left(\frac{P_{R1} + P_{R2}}{2}\right)\right]$$

The deterministic probabilities in $\mathbf{G} = \mathcal{C} = \mathcal{R} + \mathcal{M}$.
We have: In $\mathbf{G_1}$:

$$Pc_1^2 = [P_{R1} + P_{M1}/i]^2 = [P_{R1} + (1 - P_{R1})]^2 = 1^2 = 1$$
$$= DOK_1 - Chf_1 = 1$$
$$= DOK_1 + MChf_1 = 1$$
$$= Pc_1$$

In $\mathbf{G}_2$:

$$\begin{aligned} Pc_2^2 &= [P_{R2} + P_{M2}/i]^2 = [P_{R2} + (1 - P_{R2})]^2 = 1^2 = 1 \\ &= DOK_2 - Chf_2 = 1 \\ &= DOK_2 + MChf_2 = 1 \\ &= Pc_2 \end{aligned}$$

Then, in $\mathbf{G}$:

$$\begin{aligned} Pc^2 &= [P_R + P_M/i]^2 = [P_R + (1 - P_R)]^2 = 1^2 = 1 \\ &= DOK - Chf = 1 \\ &= DOK + MChf = 1 \\ &= Pc \end{aligned}$$

6.3 The deterministic cases and the *MCPP* parameters of the first model

Additionally, it is crucial and very important to mention here that if the real probabilities P_{R1} and P_{R2} are equal to one or zero, then we will return directly to the deterministic theory, which is a special nonrandom case of the probabilistic complex probability paradigm (*MCPP*) general case. Hence, this certainly proves that *MCPP* is always valid in the deterministic case or in the probabilistic and random case.

Consequently, in the deterministic situation, we will have these possible four cases:

Case 1:

$P_{R1} = 1$ and $P_{R2} = 1$, that means we are working in $\mathbf{G} = \mathbf{G_1}\ (0 \le v < c) + \mathbf{G_2}\ (c < v \le 2c)$.

$P_{M1} = i(1 - P_{R1}) = i(1 - 1) = 0$ and $P_{M2} = i(1 - P_{R2}) = i(1 - 1) = 0$.

$P_{M1}/i = i(1 - P_{R1})/i = 1 - P_{R1} = 1 - 1 = 0$ and $P_{M2}/i = i(1 - P_{R2})/i = 1 - P_{R2} = 1 - 1 = 0$.

$P_{R1} + P_{M1}/i = P_{R1} + i(1 - P_{R1})/i = 1 + 0 = 1 = P_{G1} = Pc_1$ and $P_{R2} + P_{M2}/i = P_{R2} + i(1 - P_{R2})/i = 1 + 0 = 1 = P_{G2} = Pc_2$.

Hence, $P_R = \frac{P_{R1}+P_{R2}}{2} = \frac{1+1}{2} = 1$ and $P_M = \frac{P_{M1}+P_{M2}}{2} = \frac{0+0}{2} = 0$.

So, $P_R + P_M/i = P_R + i(1 - P_R)/i = 1 + 0 = 1 = Pc$.

And $P_G = \frac{P_{G1}+P_{G2}}{2} = \frac{1+1}{2} = 1 = Pc$

$$Z_1 = P_{R1} + P_{M1} = 1 + 0 = 1$$

$$Z_2 = P_{R2} + P_{M2} = 1 + 0 = 1$$

$$Z = P_R + P_M = 1 + 0 = 1 = \frac{Z_1 + Z_2}{2} = \frac{1+1}{2} = 1$$

$$DOK_1 = |Z_1|^2 = P_{R1}^2 + (P_{M1}/i)^2 = 1^2 + 0^2 = 1$$

$$DOK_2 = |Z_2|^2 = P_{R2}^2 + (P_{M2}/i)^2 = 1^2 + 0^2 = 1$$

$$DOK = |Z|^2 = P_R^2 + (P_M/i)^2 = 1^2 + 0^2 = 1$$

$$Chf_1 = 2iP_{R1}P_{M1} = 2i \times 1 \times 0 = 0$$

$$Chf_2 = 2iP_{R2}P_{M2} = 2i \times 1 \times 0 = 0$$

$$Chf = 2iP_RP_M = 2i \times 1 \times 0 = 0$$

$$MChf_1 = |Chf_1| = -2iP_{R1}P_{M1} = -2i \times 1 \times 0 = 0$$

$$MChf_2 = |Chf_2| = -2iP_{R2}P_{M2} = -2i \times 1 \times 0 = 0$$

$$MChf = |Chf| = -2iP_RP_M = -2i \times 1 \times 0 = 0$$

$$\begin{aligned} Pc_1^2 &= (P_{R1} + P_{M1}/i)^2 = (1+0)^2 = 1^2 = 1 \\ &= DOK_1 - Chf_1 = 1 - 0 = 1 \\ &= DOK_1 + MChf_1 = 1 + 0 = 1 \\ &= Pc_1 \end{aligned}$$

$$\begin{aligned} Pc_2^2 &= (P_{R2} + P_{M2}/i)^2 = (1+0)^2 = 1^2 = 1 \\ &= DOK_2 - Chf_2 = 1 - 0 = 1 \\ &= DOK_2 + MChf_2 = 1 + 0 = 1 \\ &= Pc_2 \end{aligned}$$

$$\begin{aligned} Pc^2 &= (P_R + P_M/i)^2 = (1+0)^2 = 1^2 = 1 \\ &= DOK - Chf = 1 - 0 = 1 \\ &= DOK + MChf = 1 + 0 = 1 \\ &= Pc \end{aligned}$$

Case 2:

$P_{R1} = 1$ and $P_{R2} = 0$, that means we are working in $\mathbf{G} = \mathbf{G_1}$ $(0 \le v < c)$ alone.

$P_{M1} = i(1 - P_{R1}) = i(1-1) = 0$ and $P_{M2} = i(1 - P_{R2}) = i(1-0) = i$.

$P_{M1}/i = i(1 - P_{R1})/i = 1 - P_{R1} = 1 - 1 = 0$ and $P_{M2}/i = i(1 - P_{R2})/i = 1 - P_{R2} = 1 - 0 = 1$.

$P_{R1} + P_{M1}/i = P_{R1} + i(1 - P_{R1})/i = 1 + 0 = 1 = P_{G1} = Pc_1$ and $P_{R2} + P_{M2}/i = P_{R2} + i(1 - P_{R2})/i = 0 + 1 = 1 = P_{G2} = Pc_2$.

Hence, $P_R = \frac{P_{R1}+P_{R2}}{2} = \frac{1+0}{2} = 0.5$ and $P_M = \frac{P_{M1}+P_{M2}}{2} = \frac{0+i}{2} = 0.5i$.

So, $P_R + P_M/i = P_R + i(1 - P_R)/i = 0.5 + 0.5 = 1 = Pc$.

And $P_G = \frac{P_{G1}+P_{G2}}{2} = \frac{1+1}{2} = 1 = Pc$

$$Z_1 = P_{R1} + P_{M1} = 1 + 0 = 1$$

$$Z_2 = P_{R2} + P_{M2} = 0 + i = i$$

$$Z = P_R + P_M = 0.5 + 0.5i = \frac{Z_1 + Z_2}{2} = \frac{1+i}{2} = \frac{1}{2} + \frac{i}{2} = 0.5 + 0.5i$$

$$DOK_1 = |Z_1|^2 = P_{R1}^2 + (P_{M1}/i)^2 = 1^2 + 0^2 = 1$$

$$DOK_2 = |Z_2|^2 = P_{R2}^2 + (P_{M2}/i)^2 = 0^2 + 1^2 = 1$$

$$DOK = |Z|^2 = P_R^2 + (P_M/i)^2 = 0.5^2 + 0.5^2 = 0.5$$

$$Chf_1 = 2iP_{R1}P_{M1} = 2i \times 1 \times 0 = 0$$

$$Chf_2 = 2iP_{R2}P_{M2} = 2i \times 0 \times i = 0$$

$$Chf = 2iP_RP_M = 2i \times 0.5 \times 0.5i = -0.5$$

$$MChf_1 = |Chf_1| = -2iP_{R1}P_{M1} = -2i \times 1 \times 0 = 0$$

$$MChf_2 = |Chf_2| = -2iP_{R2}P_{M2} = -2i \times 0 \times i = 0$$

$$MChf = |Chf| = -2iP_RP_M = -2i \times 0.5 \times 0.5i = 0.5$$

$$\begin{aligned} Pc_1^2 &= (P_{R1} + P_{M1}/i)^2 = (1+0)^2 = 1^2 = 1 \\ &= DOK_1 - Chf_1 = 1 - 0 = 1 \\ &= DOK_1 + MChf_1 = 1 + 0 = 1 \\ &= Pc_1 \end{aligned}$$

$$\begin{aligned} Pc_2^2 &= (P_{R2} + P_{M2}/i)^2 = (0+1)^2 = 1^2 = 1 \\ &= DOK_2 - Chf_2 = 1 - 0 = 1 \\ &= DOK_2 + MChf_2 = 1 + 0 = 1 \\ &= Pc_2 \end{aligned}$$

$$\begin{aligned} Pc^2 &= (P_R + P_M/i)^2 = (0.5 + 0.5)^2 = 1^2 = 1 \\ &= DOK - Chf = 0.5 - (-0.5) = 0.5 + 0.5 = 1 \\ &= DOK + MChf = 0.5 + 0.5 = 1 \\ &= Pc \end{aligned}$$

These results can be understood and explained since we are considering in this case only the subluminal universe $\mathbf{G}_1$ and discarding totally the superluminal universe $\mathbf{G}_2$; hence, we are looking at one part or half of the whole picture, which is the total complex universe $\mathbf{G} = \mathbf{G}_1 + \mathbf{G}_2$. Additionally, both $\mathbf{G}_1$ and $\mathbf{G}_2$ have equal probabilities to be considered in **G** and which are $P_R = P_M/i = \frac{1}{2} = 0.5$. Moreover, and for the same reason, *DOK* which is the degree of our knowledge in the whole universe **G** is minimum and is equal to 0.5; hence, accordingly *MChf*, which measures the magnitude of chaos and ignorance in **G** is maximum and is equal to 0.5. Knowing that *Pc* in **G**, which is computed by subtracting and eliminating chaos materialized by *Chf* from the experiment and after adding the contributions of $\mathcal{M}$ to **G**, is always maintained as equal to 1 = 100%.

Case 3:

$P_{R1} = 0$ and $P_{R2} = 1$, that means we are working in $\mathbf{G} = \mathbf{G}_2$ $(c < v \leq 2c)$ alone.

$P_{M1} = i(1 - P_{R1}) = i(1 - 0) = i$ and $P_{M2} = i(1 - P_{R2}) = i(1 - 1) = 0$.

$P_{M1}/i = i(1 - P_{R1})/i = 1 - P_{R1} = 1 - 0 = 1$ and $P_{M2}/i = i(1 - P_{R2})/i = 1 - P_{R2} = 1 - 1 = 0$.

$P_{R1} + P_{M1}/i = P_{R1} + i(1 - P_{R1})/i = 0 + 1 = 1 = P_{G1} = Pc_1$ and $P_{R2} + P_{M2}/i = P_{R2} + i(1 - P_{R2})/i = 1 + 0 = 1 = P_{G2} = Pc_2$.

Hence, $P_R = \frac{P_{R1} + P_{R2}}{2} = \frac{0+1}{2} = 0.5$ and $P_M = \frac{P_{M1} + P_{M2}}{2} = \frac{i+0}{2} = 0.5i$.

So, $P_R + P_M/i = P_R + i(1 - P_R)/i = 0.5 + 0.5 = 1 = Pc$.

And $P_G = \frac{P_{G1} + P_{G2}}{2} = \frac{1+1}{2} = 1 = Pc$

$$Z_1 = P_{R1} + P_{M1} = 0 + i = i$$

$$Z_2 = P_{R2} + P_{M2} = 1 + 0 = 1$$

$$Z = P_R + P_M = 0.5 + 0.5i = \frac{Z_1 + Z_2}{2} = \frac{i+1}{2} = \frac{i}{2} + \frac{1}{2} = 0.5 + 0.5i$$

$$DOK_1 = |Z_1|^2 = P_{R1}^2 + (P_{M1}/i)^2 = 0^2 + 1^2 = 1$$

$$DOK_2 = |Z_2|^2 = P_{R2}^2 + (P_{M2}/i)^2 = 1^2 + 0^2 = 1$$

$$DOK = |Z|^2 = P_R^2 + (P_M/i)^2 = 0.5^2 + 0.5^2 = 0.5$$

$$Chf_1 = 2iP_{R1}P_{M1} = 2i \times 0 \times i = 0$$

$$Chf_2 = 2iP_{R2}P_{M2} = 2i \times 1 \times 0 = 0$$

$$Chf = 2iP_R P_M = 2i \times 0.5 \times 0.5i = -0.5$$

$$MChf_1 = |Chf_1| = -2iP_{R1}P_{M1} = -2i \times 0 \times i = 0$$

$$MChf_2 = |Chf_2| = -2iP_{R2}P_{M2} = -2i \times 1 \times 0 = 0$$

$$MChf = |Chf| = -2iP_R P_M = -2i \times 0.5 \times 0.5i = 0.5$$

$$\begin{aligned} Pc_1^2 &= (P_{R1} + P_{M1}/i)^2 = (0+1)^2 = 1^2 = 1 \\ &= DOK_1 - Chf_1 = 1 - 0 = 1 \\ &= DOK_1 + MChf_1 = 1 + 0 = 1 \\ &= Pc_1 \end{aligned}$$

$$\begin{aligned} Pc_2^2 &= (P_{R2} + P_{M2}/i)^2 = (1+0)^2 = 1^2 = 1 \\ &= DOK_2 - Chf_2 = 1 - 0 = 1 \\ &= DOK_2 + MChf_2 = 1 + 0 = 1 \\ &= Pc_2 \end{aligned}$$

$$\begin{aligned} Pc^2 &= (P_R + P_M/i)^2 = (0.5 + 0.5)^2 = 1^2 = 1 \\ &= DOK - Chf = 0.5 - (-0.5) = 0.5 + 0.5 = 1 \\ &= DOK + MChf = 0.5 + 0.5 = 1 \\ &= Pc \end{aligned}$$

These results can be understood and explained since we are considering in this case only the superluminal universe $\mathbf{G}_2$ and discarding totally the subluminal universe $\mathbf{G}_1$; hence, we are looking at one part or half of the whole picture, which is the total complex universe $\mathbf{G} = \mathbf{G}_1 + \mathbf{G}_2$. Additionally, both $\mathbf{G}_1$ and $\mathbf{G}_2$ have equal probabilities to be considered in **G** and which are $P_R = P_M/i = \frac{1}{2} = 0.5$. Moreover, and for the same reason, *DOK* which is the degree of our knowledge in the whole universe **G** is minimum and is equal to 0.5; hence, accordingly *MChf* which measures the magnitude of chaos and ignorance in **G** is maximum and is equal to 0.5. Knowing that *Pc* in **G**, which is computed by subtracting and eliminating chaos materialized by *Chf* from the experiment and after adding the contributions of $\boldsymbol{\mathcal{M}}$ to **G**, is always maintained as equal to 1 = 100%.

Case 4:

$P_{R1} = 0$ and $P_{R2} = 0$, that means that we have impossible events and experiments in the whole **G**.

$P_{M1} = i(1 - P_{R1}) = i(1 - 0) = i$ and $P_{M2} = i(1 - P_{R2}) = i(1 - 0) = i$.

$P_{M1}/i = i(1-P_{R1})/i = 1-P_{R1} = 1-0 = 1$ and $P_{M2}/i = i(1-P_{R2})/i = 1-P_{R2} = 1-0 = 1$.

$P_{R1}+P_{M1}/i = P_{R1}+i(1-P_{R1})/i = 0+1 = 1 = P_{G1} = Pc_1$ and $P_{R2}+P_{M2}/i = P_{R2}+i(1-P_{R2})/i = 0+1 = 1 = P_{G2} = Pc_2$.

Hence, $P_R = \frac{P_{R1}+P_{R2}}{2} = \frac{0+0}{2} = 0$ and $P_M = \frac{P_{M1}+P_{M2}}{2} = \frac{i+i}{2} = i$.

So, $P_R + P_M/i = P_R + i(1-P_R)/i = 0+1 = 1 = Pc$.

And $P_G = \frac{P_{G1}+P_{G2}}{2} = \frac{1+1}{2} = 1 = Pc$

$$Z_1 = P_{R1} + P_{M1} = 0 + i = i$$

$$Z_2 = P_{R2} + P_{M2} = 0 + i = i$$

$$Z = P_R + P_M = 0 + i = i = \frac{Z_1 + Z_2}{2} = \frac{i+i}{2} = \frac{2i}{2} = i$$

$$DOK_1 = |Z_1|^2 = P_{R1}^2 + (P_{M1}/i)^2 = 0^2 + 1^2 = 1$$

$$DOK_2 = |Z_2|^2 = P_{R2}^2 + (P_{M2}/i)^2 = 0^2 + 1^2 = 1$$

$$DOK = |Z|^2 = P_R^2 + (P_M/i)^2 = 0^2 + 1^2 = 1$$

$$Chf_1 = 2iP_{R1}P_{M1} = 2i \times 0 \times i = 0$$

$$Chf_2 = 2iP_{R2}P_{M2} = 2i \times 0 \times i = 0$$

$$Chf = 2iP_R P_M = 2i \times 0 \times i = 0$$

$$MChf_1 = |Chf_1| = -2iP_{R1}P_{M1} = -2i \times 0 \times i = 0$$

$$MChf_2 = |Chf_2| = -2iP_{R2}P_{M2} = -2i \times 0 \times i = 0$$

$$MChf = |Chf| = -2iP_R P_M = -2i \times 0 \times i = 0$$

$$\begin{aligned} Pc_1^2 &= (P_{R1} + P_{M1}/i)^2 = (0+1)^2 = 1^2 = 1 \\ &= DOK_1 - Chf_1 = 1 - 0 = 1 \\ &= DOK_1 + MChf_1 = 1 + 0 = 1 \\ &= Pc_1 \end{aligned}$$

$$\begin{aligned} Pc_2^2 &= (P_{R2} + P_{M2}/i)^2 = (0+1)^2 = 1^2 = 1 \\ &= DOK_2 - Chf_2 = 1 - 0 = 1 \\ &= DOK_2 + MChf_2 = 1 + 0 = 1 \\ &= Pc_2 \end{aligned}$$

$$\begin{aligned} Pc^2 &= (P_R + P_M/i)^2 = (0+1)^2 = 1^2 = 1 \\ &= DOK - Chf = 1 - 0 = 1 \\ &= DOK + MChf = 1 + 0 = 1 \\ &= Pc \end{aligned}$$

6.4 The first model simulations

We note that in the following simulations, P_{R3} is the real probability in the luminal universe $\mathbf{G_3}$ for $(v = c)$ in yellow in the simulations, where we have $\forall P_{R3} : 0 \leq P_{R3} \leq 1$

and that it will be included in the final most general model of *MCPP*. Thus, the current model is a simplified first model. The simulations from **Figures 9–12** illustrate the first model.

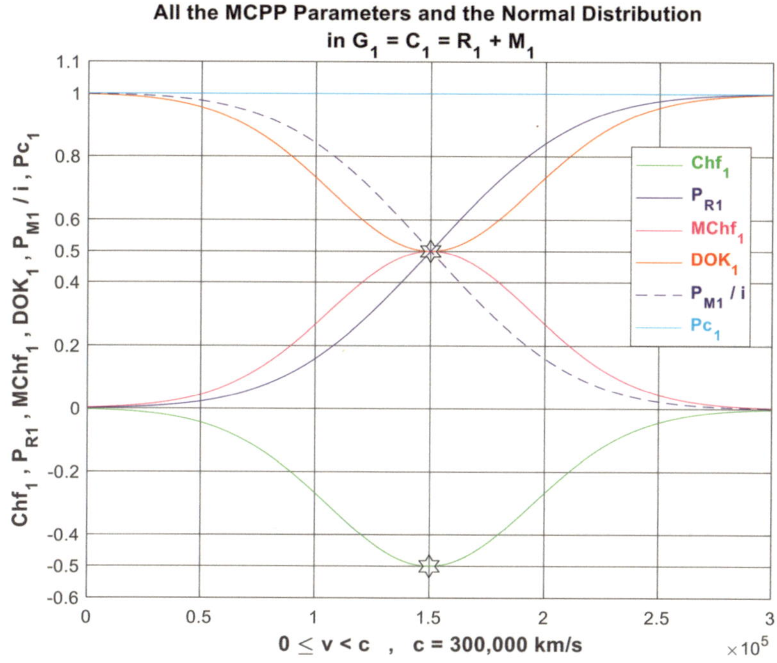

Figure 9.
The MCPP *first model parameters and the normal distribution in* G_1.

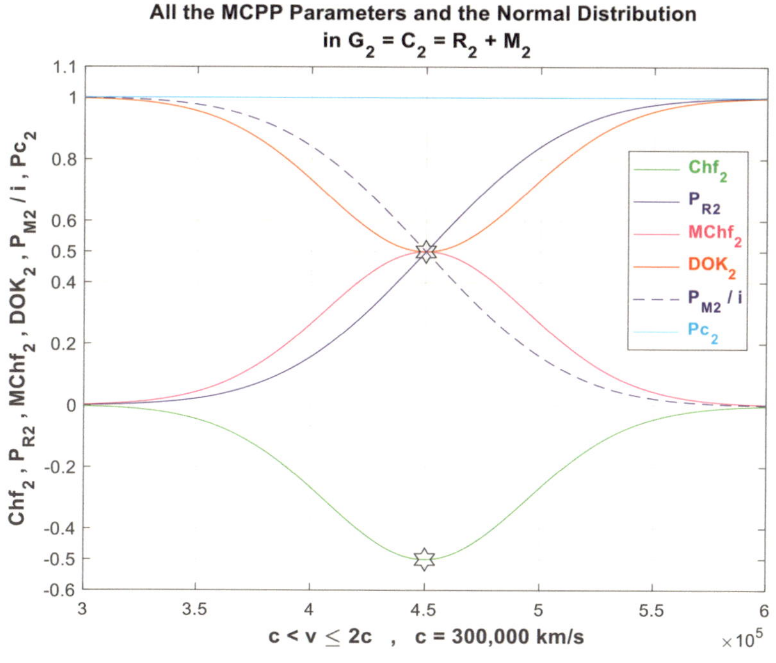

Figure 10.
The MCPP *first model parameters and the normal distribution in* G_2.

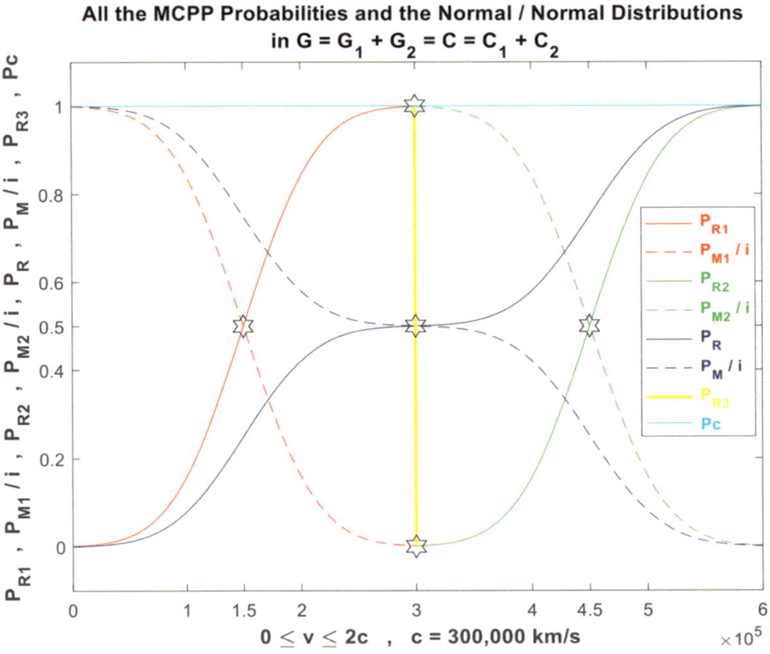

Figure 11.
The MCPP *first model probabilities and the normal/normal distributions in* **G**.

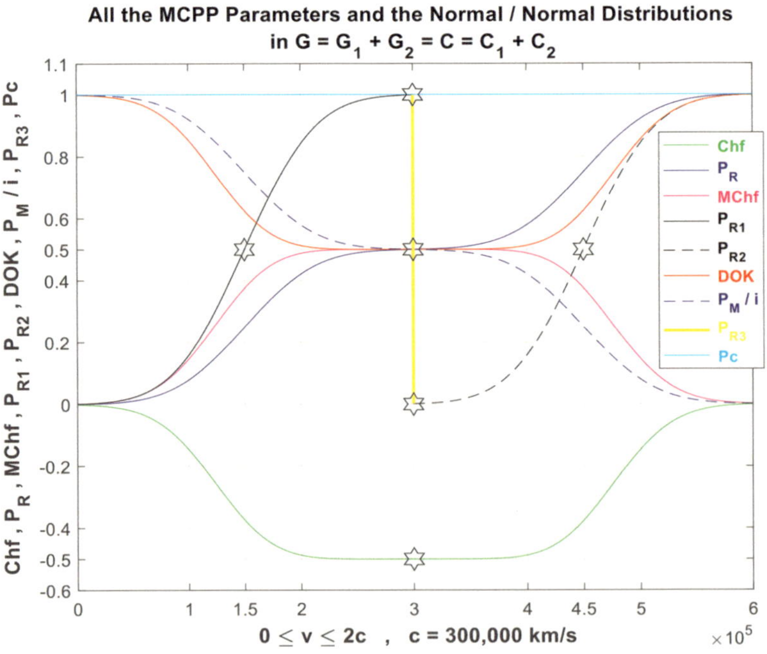

Figure 12.
The MCPP *first model parameters and the normal/normal distributions in* **G**.

7. Conclusion

In the current research work, the original extended model of eight axioms (*EKA*) of A. N. Kolmogorov was connected and applied to Metarelativity theory. Thus, a tight link between Metarelativity and the novel paradigm (*CPP*) was achieved. Consequently, the model of "Complex Probability" was more developed beyond the scope of my 21 previous research works on this topic.

Additionally, as it was proved and verified in the novel model, before the beginning of the random phenomenon simulation and at its end we have the chaotic factor (*Chf* and *MChf*) is zero and the degree of our knowledge (*DOK*) is one since the stochastic fluctuations and effects have either not started yet or they have terminated and finished their task on the probabilistic phenomenon. During the execution of the nondeterministic phenomenon and experiment, we also have $0.5 \leq DOK < 1$, $-0.5 \leq Chf < 0$, and $0 < MChf \leq 0.5$. We can see that during this entire process, we have incessantly and continually $Pc^2 = DOK - Chf = DOK + MChf = 1 = Pc$, which means that the simulation that behaved randomly and stochastically in the real probability set $\mathcal{R}$ is now certain and deterministic in the complex probability set and total universe $\mathbf{G} = \mathcal{C} = \mathcal{R} + \mathcal{M}$, and this after adding to the random experiment executed in the real probability set $\mathcal{R}$ the contributions of the imaginary probability set $\mathcal{M}$ and hence after eliminating and subtracting the chaotic factor from the degree of our knowledge as it is shown in the equation above. Furthermore, the real, imaginary, complex, and deterministic probabilities and that correspond to each value of the velocity random variable have been determined in the three probabilities sets, which are $\mathcal{R}$, $\mathcal{M}$, and $\mathbf{G} = \mathcal{C}$ by P_r, P_m, Z, and Pc, respectively. Consequently, at each value of v the novel *MCPP* parameters P_r, P_m, P_m/i, *DOK*, *Chf*, *MChf*, *Pc*, and Z are surely and perfectly predicted in the complex probabilities set and total universe $\mathbf{G} = \mathcal{C} = \mathbf{G_1} + \mathbf{G_2} + \mathbf{G_3}$ with *Pc* maintained equal to one permanently and repeatedly.

Nomenclature

EKA	Extended Kolmogorov's Axioms
CPP	Complex Probability Paradigm
MCPP	Metarelativistic Complex Probability Paradigm
i	the imaginary number where $i = \sqrt{-1}$ or $i^2 = -1$
$\mathbf{G_1}$	real universe of matter and energy = subluminal universe
$\mathbf{G_2}$	imaginary universe of matter and energy = superluminal universe or metauniverse
$\mathbf{G_3}$	luminal universe of electromagnetic waves
$\mathbf{G}$	total universe of matter and energy $= \mathbf{G_1}\ (v < c) + \mathbf{G_2}\ (v > c) + \mathbf{G_3}\ (v = c) =$ complex universe
R_1	real probabilities set in $\mathbf{G_1}(v < c)$
M_1	imaginary complementary probabilities set to R_1 in $\mathbf{G_1}\ (v < c)$
R_2	real probabilities set in $\mathbf{G_2}\ (v > c)$
M_2	imaginary complementary probabilities set to R_2 in $\mathbf{G_2}\ (v > c)$
R_3	real probabilities set in $\mathbf{G_3}\ (v = c)$
M_3	imaginary complementary probabilities set to R_3 in $\mathbf{G_3}\ (v = c)$
$\mathcal{C}_1$	$R_1 + M_1$ = complex set of probabilities in $\mathbf{G_1}\ (v < c)$
$\mathcal{C}_2$	$R_2 + M_2$ = complex set of probabilities in $\mathbf{G_2}\ (v > c)$

$\mathcal{C}_3$	$R_3 + M_3$ = complex set of probabilities in $\mathbf{G_3}$ $(v = c)$
$\mathcal{R}$	$R_1 + R_2 + R_3$ = real set of events and probabilities in **G**
$\mathcal{M}$	$M_1 + M_2 + M_3$ = imaginary set of events and probabilities in **G**
$\mathcal{C}$	complex set of events and probabilities in **G**, $\mathcal{C} = \mathcal{R} + \mathcal{M}$
P_{rob}	probability of any event
P_{R1}	probability in the real set $\mathcal{R}$ in $\mathbf{G_1}$
P_{R2}	probability in the real set $\mathcal{R}$ in $\mathbf{G_2}$
P_{R3}	probability in the real set $\mathcal{R}$ in $\mathbf{G_3}$
P_{M1}	probability in the imaginary set $\mathcal{M}$ in $\mathbf{G_1}$
P_{M2}	probability in the imaginary set $\mathcal{M}$ in $\mathbf{G_2}$
P_{M3}	probability in the imaginary set $\mathcal{M}$ in $\mathbf{G_3}$
P_R	probability in the real set $\mathcal{R}$ in **G**
P_M	probability in the imaginary set $\mathcal{M}$ in **G**
Pc_1	probability in the complex set $\mathcal{C}_1$ in $\mathbf{G_1}$
Pc_2	probability in the complex set $\mathcal{C}_2$ in $\mathbf{G_2}$
Pc_3	probability in the complex set $\mathcal{C}_3$ in $\mathbf{G_3}$
Pc	probability of a real event in $\mathcal{R}$ with its associated complementary imaginary event in $\mathcal{M}$ = probability in the complex probability set $\mathcal{C}$ in the total universe **G**
Z	complex probability number = sum of P_R and P_M = complex random vector
DOK	$= \lvert Z \rvert^2 =$ the degree of our knowledge of the random system or experiment, it is the square of the norm of Z
Chf	the chaotic factor of Z
$MChf$	the magnitude of the chaotic factor of Z
c	light velocity $\cong$ 300,000 Km/s = 3×10^8 m/s $\cong$ 186,000 miles/s in vacuum
m_{G1}	mass in the real subluminal universe $\mathbf{G_1}$ of matter
m_{G2}	mass in the imaginary superluminal universe $\mathbf{G_2}$ of matter or metamatter
m_{G3}	mass in the luminal universe $\mathbf{G_3}$
m_G	mass in the complex total universe of matter **G**
E_{G1}	energy in the real subluminal universe $\mathbf{G_1}$ of energy
E_{G2}	energy in the imaginary superluminal universe $\mathbf{G_2}$ of energy
E_{G3}	energy of the electromagnetic waves in the luminal universe $\mathbf{G_3}$ of energy
E_G	energy in the complex total universe **G** of energy
EW	electromagnetic waves in $\mathbf{G_3}$
f	frequency of the electromagnetic waves in $\mathbf{G_3}$
PDF	probability density function
CDF	cumulative probability distribution function

Author details

Abdo Abou Jaoudé
Faculty of Natural and Applied Sciences, Department of Mathematics and Statistics, Notre Dame University-Louaize, Lebanon

*Address all correspondence to: abdoaj@idm.net.lb

References

[1] Abou Jaoude A. The theory of Metarelativity: Beyond Albert Einstein's relativity. Physics International, Science Publications. 2013;**4**(2):97-109

[2] Wikipedia, the Free Encyclopedia, Special Relativity. https://en.wikipedia.org/

[3] Wikipedia, the Free Encyclopedia, General Relativity. https://en.wikipedia.org/

[4] Abou Jaoude A, El-Tawil K, Kadry S. Prediction in complex dimension using Kolmogorov's set of axioms. Journal of Mathematics and Statistics, Science Publications. 2010;**6**(2):116-124

[5] Abou Jaoude A. The Complex statistics paradigm and the law of large numbers. Journal of Mathematics and Statistics, Science Publications. 2013;**9** (4):289-304

[6] Abou Jaoude A. The theory of complex probability and the first order reliability method. Journal of Mathematics and Statistics, Science Publications. 2013;**9**(4):310-324

[7] Abou Jaoude A. Complex probability theory and prognostic. Journal of Mathematics and Statistics, Science Publications. 2014;**10**(1):1-24

[8] Abou Jaoude A. The complex probability paradigm and analytic linear prognostic for vehicle suspension systems. American Journal of Engineering and Applied Sciences, Science Publications. 2015;**8**(1):147-175

[9] Abou Jaoude A. The paradigm of complex probability and the Brownian motion. Systems Science and Control Engineering, Taylor and Francis Publishers. 2015;**3**(1):478-503

[10] Abou Jaoude A. The paradigm of complex probability and Chebyshev's inequality. Systems Science and Control Engineering, Taylor and Francis Publishers. 2016;**4**(1):99-137

[11] Abou Jaoude A. The paradigm of complex probability and analytic nonlinear prognostic for vehicle suspension systems. Systems Science and Control Engineering. Taylor and Francis Publishers. 2016;**4**(1):99-137

[12] Abou Jaoude A. The paradigm of complex probability and analytic linear prognostic for unburied petrochemical pipelines. Systems Science and Control Engineering. Taylor and Francis Publishers. 2017;**5**(1):178-214

[13] Abou Jaoude A. The paradigm of complex probability and Claude Shannon's information theory. Systems Science and Control Engineering. Taylor and Francis Publishers. 2017;**5**(1): 380-425

[14] Abou Jaoude A. The paradigm of complex probability and analytic nonlinear prognostic for unburied petrochemical pipelines. Systems Science and Control Engineering. 2017;**5**(1): 495-534

[15] Abou Jaoude A. The paradigm of complex probability and Ludwig Boltzmann's entropy. Systems Science and Control Engineering. Taylor and Francis Publishers. 2018;**6**(1):108-149

[16] Abou Jaoude A. The paradigm of complex probability and Monte Carlo methods. Systems Science and Control Engineering, Taylor and Francis Publishers. 2019;**7**(1):407-451

[17] Abou Jaoude A. Analytic prognostic in the linear damage case applied to

buried petrochemical pipelines and the complex probability paradigm. Fault Detection, Diagnosis and Prognosis, IntechOpen. 2020;**1**(5):65-103. DOI: 10.5772/intechopen.90157

[18] Abou Jaoude A. The Monte Carlo techniques and the complex probability paradigm. Forecasting in Mathematics - Recent Advances, New Perspectives and Applications. IntechOpen. 2020;**1**(1):1-29. DOI: 10.5772/intechopen.93048

[19] Abou Jaoude A. The paradigm of complex probability and prognostic using FORM. London Journal of Research in Science: Natural and Formal (LJRS). 2020;**20**(4):1-65. London Journals Press, Print ISSN: 2631-8490, Online ISSN: 2631-8504. DOI: 10.17472/LJRS

[20] Abou Jaoude A. The paradigm of complex probability and the central limit theorem. London Journal of Research in Science: Natural and Formal (LJRS). 2020;**20**(5):1-57. London Journals Press, Print ISSN: 2631-8490, Online ISSN: 2631-8504. DOI: 10.17472/LJRS

[21] Abou Jaoude A. The paradigm of complex probability and Thomas Bayes' theorem. The Monte Carlo Methods - Recent Advances, New Perspectives and Applications. IntechOpen. 2021;**1**(1):1-44. ISBN: 978-1-83968-760-0. DOI: 10.5772/intechopen.98340

[22] Abou Jaoude A. The paradigm of complex probability and Isaac Newton's classical mechanics: On the Foundation of Statistical Physics. The Monte Carlo Methods - Recent Advances, New Perspectives and Applications. IntechOpen. 2021;**1**(2):45-116. ISBN: 978-1-83968-760-0. DOI: 10.5772/intechopen.98341

[23] Abou Jaoude A. The Paradigm of Complex Probability and Quantum Mechanics: The Infinite Potential Well Problem - the Position Wave Function. Applied Probability Theory - New Perspectives, Recent Advances and Trends. London, UK: IntechOpen; 2022;**1**(1):1-44. DOI: 10.5772/intechopen.107300

[24] Abou Jaoude A. The paradigm of complex probability and quantum mechanics: The infinite potential well problem – The momentum Wavefunction and the Wavefunction entropies. Applied Probability Theory – New Perspectives, Recent Advances and Trends. IntechOpen. 2022;**1**(2):45-88. DOI: 10.5772/intechopen.107665

[25] Benton W. Probability, Encyclopedia Britannica. Vol. 18. Chicago: Encyclopedia Britannica Inc.; 1966. pp. 570-574

[26] Benton W. Mathematical Probability, Encyclopedia Britannica. Vol. 18. Chicago: Encyclopedia Britannica Inc.; 1966. pp. 574-579

[27] Feller W. An Introduction to Probability Theory and its Applications. 3rd ed. New York: Wiley; 1968

[28] Walpole R, Myers R, Myers S, Ye K. Probability and Statistics for Engineers and Scientists. 7th ed. New Jersey: Prentice Hall; 2002

[29] Freund JE. Introduction to Probability. New York: Dover Publications; 1973

[30] Abou Jaoude A. The Computer Simulation of Monté Carlo Methods and Random Phenomena. United Kingdom: Cambridge Scholars Publishing; 2019

[31] Abou Jaoude A. The Analysis of Selected Algorithms for the Stochastic

Paradigm. United Kingdom: Cambridge Scholars Publishing; 2019

[32] Abou Jaoude A. The Analysis of Selected Algorithms for the Statistical Paradigm. Vol. 1. The Republic of Moldova: Generis Publishing; 2021

[33] Abou Jaoude A. The Analysis of Selected Algorithms for the Statistical Paradigm. Vol. 2. The Republic of Moldova: Generis Publishing; 2021

[34] Abou Jaoude A. Forecasting in Mathematics - Recent Advances, New Perspectives and Applications. London, UK, London, United Kingdom: IntechOpen; 2021. ISBN: 978-1-83880-825-9

[35] Abou Jaoude A. The Monte Carlo Methods - Recent Advances, New Perspectives and Applications. London, UK, London, United Kingdom: IntechOpen; 2022. ISBN: 978-1-83968-760-0

[36] Abou Jaoude A. Applied Probability Theory – New Perspectives, Recent Advances and Trends. London, UK, London, United Kingdom: IntechOpen; 2023. ISBN: 978-1-83768-296-6

[37] Abou Jaoude A. Ph. D. Thesis in Applied Mathematics: Numerical Methods and Algorithms for Applied Mathematicians. Spain: Bircham International University; 2004. http://www.bircham.edu

[38] Abou Jaoude A. Ph. D. Thesis in Computer Science: Computer Simulation of Monté Carlo Methods and Random Phenomena. Spain: Bircham International University; 2005. http://www.bircham.edu

[39] Abou Jaoude A. Ph. D. Thesis in Applied Statistics and Probability: Analysis and Algorithms for the Statistical and Stochastic Paradigm. Spain: Bircham International University; 2007. http://www.bircham.edu

[40] Wikipedia, the Free Encyclopedia, Dark Matter. https://en.wikipedia.org/

[41] Wikipedia, the Free Encyclopedia, Dark Energy. https://en.wikipedia.org/

[42] Wikipedia, the Free Encyclopedia, Lambda-CDM Model. https://en.wikipedia.org/

[43] Wikipedia, the Free Encyclopedia, Relativity: The Special and the General Theory. https://en.wikipedia.org/

Chapter 2

The Paradigm of Complex Probability and the Theory of Metarelativity: The General Model and Some Consequences of MCPP

Abdo Abou Jaoudé

Abstract

Calculating probabilities is a crucial task of classical probability theory. Adding supplementary dimensions to nondeterministic experiments will yield a deterministic expression of the theory of probability. This is the novel and original idea at the foundation of my complex probability paradigm. As a matter of fact, probability theory is a stochastic system of axioms in its essence; that means that the phenomena outputs are due to randomness and chance. By adding novel imaginary dimensions to the nondeterministic phenomenon happening in the set $\mathcal{R}$ will lead to a deterministic phenomenon and thus a stochastic experiment will have a certain output in the complex probability set and total universe $\mathbf{G} = \mathcal{C}$. If the chaotic experiment becomes completely predictable, then we will be fully capable to predict the output of random events that arise in the real world in all stochastic processes. Accordingly, the task that has been achieved here was to extend the random real probabilities set $\mathcal{R}$ to the deterministic complex probabilities set and total universe $\mathbf{G} = \mathcal{C} = \mathcal{R} + \mathcal{M}$ and this by incorporating the contributions of the set $\mathcal{M}$, which is the complementary imaginary set of probabilities to the set $\mathcal{R}$. Consequently, since this extension reveals to be successful, then an innovative paradigm of stochastic sciences and prognostic was put forward in which all nondeterministic phenomena in $\mathcal{R}$ was expressed deterministically in $\mathcal{C}$. This paradigm was initiated and elaborated in my previous 21 publications. Furthermore, this model will be linked to my theory of Metarelativity, which takes into consideration faster-than-light matter and energy. This is what I named "The Metarelativistic Complex Probability Paradigm (*MCPP*)," which will be developed in the present two chapters 1 and 2.

Keywords: degree of our knowledge, complex random vector, chaotic factor, probability norm, complex probability set $\mathcal{C}$, imaginary number, imaginary dimensions, metarelativistic transformations, superluminal velocities, metaparticles, metamatter, dark matter, dark energy, metaenergy, metaentropy, universe $\mathbf{G_1}$, metauniverse $\mathbf{G_2}$, luminal universe $\mathbf{G_3}$, the total universe $\mathbf{G}$

1. The metarelativistic complex probability paradigm (MCPP): a more general second model

In this section, we will develop the second more general model of *MCPP* with all its parameters [1–42].

1.1 The real and imaginary probabilities

Here, and in this second *MCPP* model, v_1 is always the velocity of a body in R_1 with $0 \le v_1 < c$ and is a random variable that follows the normal distribution: $N(\overline{v}_1 = c/2, \sigma_{v1} = c/6)$ where $\overline{v}_1$ is the mean or the expectation of this symmetric normal probability distribution of v_1 or $PDF_1(v_1)$ and σ_{v1} is its corresponding standard deviation. And v_2 is also the velocity of a body in R_2 with $c < v_2 \le nc$ and is a random variable that follows the normal distribution: $N(\overline{v}_2 = (n+1)c/2, \sigma_{v2} = (n-1)c/6)$ for a determined and fixed value of n such that $\forall n, n \in \mathbb{R}^+ : n > 1 \Leftrightarrow n \in (1, +\infty)$ and where $\overline{v}_2$ is the mean or the expectation of this symmetric normal probability distribution of v_2 or $PDF_2(v_2)$ and σ_{v2} is its corresponding standard deviation.

First, we will define and calculate the real and imaginary probabilities in the universes R_1, R_2, M_1, and M_2 in the second model of *MCPP* as follows:

$$P_{R1} = P_{rob}(0 \le V \le v_1) = CDF_1(0 \le V \le v_1) = \int_0^{v1} PDF_1(v)dv = \int_0^{v1} N(\overline{v} = c/2, \sigma_v = c/6)dv.$$

So, if $v_1 < 0 \Rightarrow P_{R1} = P_{rob}(V < 0) = CDF_1(V < 0) = 0$.

If $v_1 = 0 \Rightarrow P_{R1} = P_{rob}(V \le 0) = CDF_1(V \le 0) = \int_0^0 PDF_1(v)dv = \int_0^0 N(\overline{v} = c/2, \sigma_v = c/6)dv = 0$.

If $v_1 = \overline{v}_1 = c/2 \Rightarrow P_{R1} = P_{rob}(0 \le V \le c/2) = CDF_1(0 \le V \le c/2) = \int_0^{c/2} PDF_1(v)dv = 0.5$.

If $v_1 \to c^- \Rightarrow P_{R1} \to P_{rob}(0 \le V < c) = CDF_1(0 \le V < c) = \int_0^c PDF_1(v)dv = \int_0^c N(\overline{v} = c/2, \sigma_v = c/6)dv = 1$.

If $v_1 > c \Rightarrow P_{R1} = P_{rob}(V > c) = CDF_1(V > c) = \int_0^{v1} PDF_1(v)dv = \left\{ \int_0^c PDF_1(v)dv + \int_c^{v1} PDF_1(v)dv \right\} = (1 + 0) = 1$.

And we have for the second real probability:

$$\begin{aligned} P_{R2} &= P_{rob}(c < V \le v_2) = CDF_2(c < V \le v_2) \\ &= \int_c^{v2} PDF_2(v)dv = \int_c^{v2} N(\overline{v} = (n+1)c/2, \sigma_v = (n-1)c/6)dv \end{aligned}$$

So, if $v_2 < c \Rightarrow P_{R2} = P_{rob}(V < c) = CDF_2(V < c) = 0$.

If $v_2 \to c^+ \Rightarrow P_{R2} \to \int_c^c PDF_2(v)dv = \int_c^c N(\overline{v} = (n+1)c/2, \sigma_v = (n-1)c/6)dv = 0$.

If $v_2 = \overline{v}_2 = (n+1)c/2 \Rightarrow P_{R2} = P_{rob}(c < V \le (n+1)c/2) =$
$CDF_2(c < V \le (n+1)c/2) = \int_c^{(n+1)c/2} PDF_2(v)dv = 0.5.$

If $v_2 = nc$

$$\Rightarrow P_{R2} = P_{rob}(c < V \le nc) = CDF_2(c < V \le nc) = \int_c^{nc} PDF_2(v)dv = \int_c^{nc} N(\overline{v} = (n+1)c/2, \sigma_v = (n-1)c/6)dv = 1$$

If $v_2 > nc$

$$\Rightarrow P_{R2} = P_{rob}(V > nc) = CDF_2(V > nc) = \int_c^{v2} PDF_2(v)dv = \left\{ \int_c^{nc} PDF_2(v)dv + \int_{nc}^{v2} PDF_2(v)dv \right\} = (1+0) = 1$$

Moreover, the first imaginary probability is:

$$P_{M1} = i(1 - P_{R1}) = i[1 - P_{rob}(0 \le V \le v_1)] = i[1 - CDF_1(0 \le V \le v_1)] = iCDF_1(v_1 < V < c) = i\left[1 - \int_0^{v1} PDF_1(v)dv\right] = i\int_{v1}^{c} PDF_1(v)dv = i\int_{v1}^{c} N(\overline{v} = c/2, \sigma_v = c/6)dv$$

So, if $v_1 < 0 \Rightarrow$
$P_{M1} = i\int_{v1}^{c} PDF_1(v)dv = i\left\{\int_{v1}^{0} PDF_1(v)dv + \int_0^c PDF_1(v)dv\right\} = i(0+1) = i \Rightarrow P_{M1}/i = 1.$
If $v_1 = 0$
$\Rightarrow P_{M1} = i[1 - P_{rob}(V \le 0)] = i[1 - CDF_1(V \le 0)] = i(1-0) = i \Rightarrow P_{M1}/i = 1.$
If $v_1 = \overline{v}_1 = c/2$

$$\Rightarrow P_{M1} = i[1 - P_{rob}(0 \le V \le c/2)] = i[1 - CDF_1(0 \le V \le c/2)] = i\left[1 - \int_0^{c/2} PDF_1(v)dv\right] = i\int_{c/2}^{c} PDF_1(v)dv = i(1-0.5) = 0.5i \Rightarrow P_{M1}/i = 0.5$$

If $v_1 \to c^- \Rightarrow P_{M1} \to i[1 - P_{rob}(0 \le V < c)] = i[1 - CDF_1(0 \le V < c)] =$
$i\left[1 - \int_0^c PDF_1(v)dv\right] = i(1-1) = 0 \Rightarrow P_{M1}/i \to 0$
If $v_1 > c \Rightarrow P_{M1} = 0 \Rightarrow P_{M1}/i = 0.$
And we have for the second imaginary probability:

$$P_{M2} = i(1 - P_{R2}) = i[1 - P_{rob}(c < V \le v_2)] = i[1 - CDF_2(c < V \le v_2)] = iCDF_2(v_2 < V \le nc) = i\left[1 - \int_c^{v2} PDF_2(v)dv\right] = i\int_{v2}^{nc} PDF_2(v)dv = i\int_{v2}^{nc} N(\overline{v} = (n+1)c/2, \sigma_v = (n-1)c/6)dv$$

So, if $v_2 < c \Rightarrow P_{M2} = i\int_{v2}^{nc} PDF_2(v)dv = i\left\{\int_{v2}^{c} PDF_2(v)dv + \int_{c}^{nc} PDF_2(v)dv\right\} =$
$i(0+1) = i \Rightarrow P_{M2}/i = 1.$
If $v_2 \to c^+$

$$\Rightarrow P_{M2} \to i[1 - P_{rob}(V \leq v_2)] = iP_{rob}(c < V \leq nc) = iCDF_2(c < V \leq nc) = i \times 1 = i$$
$$= i[1 - CDF_2(V < c)] = i(1-0) = i$$
$$\Rightarrow P_{M2}/i \to 1$$

If $v_2 = \overline{v}_2 = (n+1)c/2$

$$\Rightarrow P_{M2} = i[1 - P_{rob}(c < V \leq (n+1)c/2)] = i[1 - CDF_2(c < V \leq (n+1)c/2)]$$
$$= i\left[1 - \int_{c}^{(n+1)c/2} PDF_2(v)dv\right] = i \times \int_{(n+1)c/2}^{nc} PDF_2(v)dv = i(1-0.5) = 0.5i \Rightarrow P_{M2}/i = 0.5$$

If $v_2 = nc \Rightarrow P_{M2} = i[1 - P_{rob}(c < V \leq nc)] = i[1 - CDF_2(c < V \leq nc)] =$

$$i\left[1 - \int_{c}^{nc} PDF_2(v)dv\right] = i(1-1) = 0$$

$$\Rightarrow P_{M2}/i = 0$$

If $v_2 > nc \Rightarrow P_{M2} = 0 \Rightarrow P_{M2}/i = 0.$
Additionally, we have $\mathcal{R} = R_1(0 \leq v < c) + R_2(c < v \leq nc), \forall n, n \in \mathbb{R}^+ : n > 1 \Leftrightarrow n \in (1, +\infty).$
Now, let $P_R = \frac{P_{R1}+P_{R2}}{2}$ and it is equal to half of the sum of the cumulative probability that $0 \leq V \leq v_1$ in R_1 and the cumulative probability that $c < V \leq v_2$ in R_2.

$$\Rightarrow P_R = \frac{CDF_1(0 \leq V \leq v_1) + CDF_2(c < V \leq v_2)}{2}$$
$$= \frac{1}{2}\left\{\int_{0}^{v1} PDF_1(v)dv + \int_{c}^{v2} PDF_2(v)dv\right\}$$
$$= \frac{1}{2}\left\{\int_{0}^{v1} N(\overline{v} = c/2, \sigma_v = c/6)dv + \int_{c}^{v2} N(\overline{v} = (n+1)c/2, \sigma_v = (n-1)c/6)dv\right\}$$

Hence, we have in $\mathbf{G} = \mathcal{C} = \mathcal{R} + \mathcal{M} = \mathbf{G_1} + \mathbf{G_2}$: $0 \leq v \leq nc$ with $v \neq c$.
So, if $0 \leq v < c \Rightarrow P_{R1} = P_{rob}(0 \leq V \leq v) = CDF_1(0 \leq V \leq v).$
And $P_{R2} = P_{rob}(V \leq v) = P_{rob}(V < c) = CDF_2(V < c) = 0$

$$\Rightarrow P_R = \frac{CDF_1(0 \leq V \leq v) + 0}{2} = \frac{CDF_1(0 \leq V \leq v)}{2} = \frac{P_{R1}}{2}$$

Therefore, we say here that we are working in the real probability universe $\mathcal{R} = R_1$ alone.
And if $c < v \leq nc \Rightarrow P_{R1} = P_{rob}(V > c) = CDF_1(V > c) = 1.$

And $P_{R2} = P_{rob}(c < V \le v) = CDF_2(c < V \le v)$

$$\Rightarrow P_R = \frac{1 + CDF_2(c < V \le v)}{2} = \frac{1 + P_{R2}}{2}$$

Therefore, we say here that we are working in the real probability universe $\mathcal{R} = R_2$ alone.

And if $0 \le v \le nc$ with $v \ne c \Rightarrow P_{R1} = P_{rob}(0 \le V \le v) = CDF_1(0 \le V \le v)$.

And $P_{R2} = P_{rob}(c < V \le v) = CDF_2(c < V \le v)$

$$\Rightarrow P_R = \frac{CDF_1(0 \le V \le v) + CDF_2(c < V \le v)}{2} = \frac{P_{R1} + P_{R2}}{2}$$

Therefore, we say here that we are working in the real probability universe $\mathcal{R} = R_1 + R_2$.

And consequently, we can deduce from the above the real probability in the probability universe $\mathcal{R} = R_1 + R_2$ for special velocity cases as follows:

if $v < 0 \Rightarrow P_R = \frac{CDF_1(V<0)}{2} = \frac{0}{2} = 0$.

if $v = c/2 \Rightarrow P_R = \frac{CDF_1(0 \le V \le c/2) + CDF_2(V<c)}{2} = \frac{0.5+0}{2} = 0.25$.

if $v \to c^- \Rightarrow P_R \to \frac{CDF_1(0 \le V < c) + CDF_2(V<c)}{2} = \frac{1+0}{2} = 0.5$.

if $v = (n+1)c/2 \Rightarrow P_R = \frac{CDF_1(0 \le V < c) + CDF_2(c < V \le (n+1)c/2)}{2} = \frac{1+0.5}{2} = 0.75$.

if $v = nc \Rightarrow P_R = \frac{CDF_1(0 \le V < c) + CDF_2(c < V \le nc)}{2} = \frac{1+1}{2} = 1$.

Furthermore, we have $\mathcal{M} = M_1(0 \le v < c) + M_2(c < v \le nc)$, $\forall n, n \in \mathbb{R}^+ : n > 1 \Leftrightarrow n \in (1, +\infty)$.

Now, let $P_M = \frac{P_{M1} + P_{M2}}{2}$ and it is equal to half of the sum of the complement of the cumulative probability that $0 \le V \le v_1$ in M_1 and the complement of the cumulative probability that $c < V \le v_2$ in M_2.

$$\Rightarrow P_M = \frac{i(1 - P_{R1}) + i(1 - P_{R2})}{2} = \frac{2i - i(P_{R1} + P_{R2})}{2} = i - \frac{i(P_{R1} + P_{R2})}{2} = i\left[1 - \frac{(P_{R1} + P_{R2})}{2}\right] = i(1 - P_R)$$

$$\Rightarrow P_M = \frac{i[1 - CDF_1(0 \le V \le v_1)] + i[1 - CDF_2(c < V \le v_2)]}{2}$$

$$= \frac{i}{2}\left\{\int_0^{v1} [1 - PDF_1(v)]dv + \int_c^{v2} [1 - PDF_2(v)]dv\right\}$$

$$= \frac{i}{2}\left\{\int_0^{v1} [1 - N(\overline{v} = c/2, \sigma_v = c/6)]dv + \int_c^{v2} [1 - N(\overline{v} = (n+1)c/2, \sigma_v = (n-1)c/6)]dv\right\}$$

$$= \frac{i}{2}\left\{\int_{v1}^{c} N(\overline{v} = c/2, \sigma_v = c/6)dv + \int_{v2}^{nc} N(\overline{v} = (n+1)c/2, \sigma_v = (n-1)c/6)dv\right\}$$

Hence, we have in $\mathbf{G} = \mathcal{C} = \mathcal{R} + \mathcal{M} = \mathbf{G_1} + \mathbf{G_2}$: $0 \le v \le nc$ with $v \ne c$.

So, if $0 \le v < c \Rightarrow P_{M1} = i[1 - P_{rob}(0 \le V \le v)] = i[1 - CDF_1(0 \le V \le v)]$.

And $P_{M2} = i[1 - P_{rob}(0 \le V \le v)] = i[1 - CDF_2(V < c)] = i(1 - 0) = i$

$$\Rightarrow P_M = \frac{i[1 - CDF_1(0 \le V \le v)] + i}{2} = \frac{i + P_{M1}}{2} = i\left[1 - \frac{P_{R1}}{2}\right]$$

Therefore, we say here that we are working in the imaginary probability universe $\mathcal{M} = M_1$ alone.

And if $c < v \le nc \Rightarrow P_{M1} = i[1 - P_{rob}(V > c)] = i[1 - CDF_1(V > c)] = i(1-1) = 0$.

And $P_{M2} = i[1 - P_{rob}(c < V \le v)] = i[1 - CDF_2(c < V \le v)]$

$$\Rightarrow P_M = \frac{0 + i[1 - CDF_2(c < V \le v)]}{2} = \frac{P_{M2}}{2} = i\left[\frac{1 - P_{R2}}{2}\right]$$

Therefore, we say here that we are working in the imaginary probability universe $\mathcal{M} = M_2$ alone.

And if

$0 \le v \le nc$ with $v \ne c \Rightarrow P_{M1} = i[1 - P_{rob}(0 \le V < v)] = i[1 - CDF_1(0 \le V < v)]$.

And $P_{M2} = i[1 - P_{rob}(c < V \le v)] = i[1 - CDF_2(c < V \le v)] \Rightarrow$

$P_M = \frac{i[1 - CDF_1(0 \le V \le v)] + i[1 - CDF_2(c < V \le v)]}{2} = \frac{P_{M1} + P_{M2}}{2} = i\left[1 - \frac{P_{R1} + P_{R2}}{2}\right] = i[1 - P_R]$.

Therefore, we say here that we are working in the imaginary probability universe $\mathcal{M} = M_1 + M_2$.

And consequently, we can deduce from the above the imaginary probability in the probability universe $\mathcal{M} = M_1 + M_2$ for special velocity cases as follows:

if $v < 0 \Rightarrow P_M = i\left[1 - \frac{CDF_1(V < 0)}{2}\right] = i\left[1 - \frac{0}{2}\right] = i \Rightarrow P_M/i = 1$.

if $v = c/2 \Rightarrow P_M = i\left[1 - \frac{CDF_1(0 \le V \le c/2) + CDF_2(V < c)}{2}\right] = i\left[1 - \frac{0.5 + 0}{2}\right] = 0.75i$

$$\Rightarrow P_M/i = 0.75$$

if $v \to c^- \Rightarrow P_M \to i\left[1 - \frac{CDF_1(0 \le V < c) + CDF_2(V < c)}{2}\right] = i\left[1 - \frac{1+0}{2}\right] = 0.5i \Rightarrow P_M/i \to 0.5$.

if $v = (n+1)c/2$

$$\Rightarrow P_M = i\left[1 - \frac{CDF_1(0 \le V < c) + CDF_2(c < V \le (n+1)c/2)}{2}\right] = i\left[1 - \frac{1 + 0.5}{2}\right] = 0.25i$$

$$\Rightarrow P_M/i = 0.25$$

if $v = nc \Rightarrow P_M = i\left[1 - \frac{CDF_1(0 \le V < c) + CDF_2(c < V \le nc)}{2}\right] = i\left[1 - \frac{1+1}{2}\right] = i(1-1) = 0$

$$\Rightarrow P_M/i = 0$$

Therefore, for any value of $0 \le v \le nc$ with $v \ne c$, we can write without any confusion that:

$P_{M1} = i(1 - P_{R1})$ and $P_{R1} = 1 - P_{M1}/i$; hence, M_1 is the imaginary complementary probability universe to the real probability universe R_1.

And $P_{M2} = i(1 - P_{R2})$ and $P_{R2} = 1 - P_{M2}/i$; hence, M_2 is the imaginary complementary probability universe to the real probability universe R_2.

Moreover, in all cases and for any value of $v: 0 \le v \le nc$ with $v \ne c$, we have:

$P_R = \frac{P_{R1} + P_{R2}}{2}$ where $\mathcal{R} = R_1 + R_2$.

And $P_M = \frac{P_{M1} + P_{M2}}{2}$ where $\mathcal{M} = M_1 + M_2$.

We can check that:

$$P_M = \frac{i(1 - P_{R1}) + i(1 - P_{R2})}{2} = \frac{2i - i(P_{R1} + P_{R2})}{2} = i - \frac{i(P_{R1} + P_{R2})}{2} = i\left[1 - \frac{(P_{R1} + P_{R2})}{2}\right] = i(1 - P_R)$$

Hence, $\mathcal{M}$ is the imaginary complementary probability universe to the real probability universe $\mathcal{R}$.

Moreover, we have in $\mathbf{G} = \mathcal{C} = \mathcal{R} + \mathcal{M}$, where $0 \le v \le nc$ with $v \ne c$,

$$\forall n, n \in \mathbb{R}^+ : n > 1 \Leftrightarrow n \in (1, +\infty).$$

Then, $\mathcal{C} = (R_1 + R_2) + (M_1 + M_2) = (R_1 + M_1) + (R_2 + M_2) = \mathcal{C}_1 + \mathcal{C}_2$.

In fact, in $\mathcal{C}_1$ we have: $Pc_1 = P_{R1} + P_{M1}/i = P_{R1} + (1 - P_{R1}) = 1$.

And, in $\mathcal{C}_2$ we have: $Pc_2 = P_{R2} + P_{M2}/i = P_{R2} + (1 - P_{R2}) = 1$.

And, in $\mathcal{C}$ we have:

$$\begin{aligned} Pc = P_R + P_M/i &= \frac{P_{R1} + P_{R2}}{2} + \left[\frac{P_{M1} + P_{M2}}{2}\right]/i = \frac{P_{R1} + P_{R2}}{2} + \left[\frac{i(1 - P_{R1}) + i(1 - P_{R2})}{2}\right]/i \\ &= \frac{P_{R1} + P_{R2}}{2} + \frac{(1 - P_{R1}) + (1 - P_{R2})}{2} = \frac{P_{R1} + P_{R2}}{2} + 1 - \frac{P_{R1} + P_{R2}}{2} \\ &= 1 \end{aligned}$$

We can calculate Pc in another way as follows:

$$\begin{aligned} Pc = P_R + P_M/i &= \frac{P_{R1} + P_{R2}}{2} + \left[\frac{P_{M1} + P_{M2}}{2}\right]/i = \frac{P_{R1} + P_{M1}/i}{2} + \frac{P_{R2} + P_{M2}/i}{2} = \frac{Pc_1}{2} + \frac{Pc_2}{2} \\ &= \frac{Pc_1 + Pc_2}{2} = \frac{1+1}{2} = 1 \end{aligned}$$

Consequently: $Pc = Pc_1 = Pc_2 = 1$, in accordance with *CPP* axioms.

Furthermore, we can state now and affirm finally that in this second model:

$\mathbf{G} = \mathbf{G_1}\ (0 \le v < c) + \mathbf{G_2}\ (c < v \le nc)$ that means that the total universe **G** is the sum of the real subluminal universe $\mathbf{G_1}$ and the imaginary superluminal universe or metauniverse $\mathbf{G_2}$.

1. The real subluminal universe $\mathbf{G_1}$ corresponds to the complex probability universe $\mathcal{C}_1$, which is also subluminal; hence, $\mathbf{G_1} = \mathcal{C}_1 = R_1 + M_1$ with $(0 \le v < c)$.

2. And the imaginary superluminal universe $\mathbf{G_2}$ or metauniverse corresponds to the complex probability universe $\mathcal{C}_2$, which is also superluminal; hence, $\mathbf{G_2} = \mathcal{C}_2 = R_2 + M_2$ with $(c < v \le nc)$.

Therefore,

$$P_{G1} = Pc_1 = P_{R1} + P_{M1}/i = P_{R1} + (1 - P_{R1}) = 1 \text{ and}$$

$$P_{G2} = Pc_2 = P_{R2} + P_{M2}/i = P_{R2} + (1 - P_{R2}) = 1.$$

Consequently, the complex total universe $\mathbf{G} = \mathbf{G_1}\ (0 \le v < c) + \mathbf{G_2}\ (c < v \le nc)$, which is the sum of the universe and the metauniverse, corresponds to the complex probability universe $\mathcal{C}$ having:

$\mathbf{G} = \mathcal{C} = \mathcal{R} + \mathcal{M} = (R_1 + R_2) + (M_1 + M_2) = (R_1 + M_1) + (R_2 + M_2)$
$= \mathcal{C}_1\ (0 \le v < c) + \mathcal{C}_2\ (c < v \le nc) = \mathbf{G_1}\ (0 \le v < c) + \mathbf{G_2}\ (c < v \le nc)$ with $0 \le v \le nc$ and $v \ne c$,

hence:

$$P_G = Pc = \frac{P_{G1} + P_{G2}}{2} = \frac{Pc_1 + Pc_2}{2} = \frac{1+1}{2} = 1$$

Consequently: $Pc = 1$, in accordance with *CPP* axioms.

Thus, we can conclude that, by adding the complementary imaginary probabilities universes M_1,M_2 and $\mathcal{M}$ to the real probabilities universes R_1,R_2 and $\mathcal{R}$ then all random phenomena in the complex probabilities' universes $\mathcal{C}_1$, $\mathcal{C}_2$, and $\mathcal{C}$, and hence in the subluminal universe $\mathbf{G_1}$, in the superluminal universe $\mathbf{G_2}$, and in the total and complex universe $\mathbf{G}$, become absolutely and perfectly deterministic with probabilities expressed totally as follows:

$Pc = Pc_1 = Pc_2 = 1$ and $P_G = P_{G1} = P_{G2} = 1$.

1.2 The *MCPP* parameters of the second model

The *MCPP* parameters in this second model are similar to those of the first model and this is done by including the probabilities P_R and P_M corresponding to the second model.

1.3 The deterministic cases and the *MCPP* parameters of the second model

The deterministic cases in this second model are similar to those of the first model and this is done by taking into consideration the probabilities P_R and P_M pertaining and corresponding to the second model.

1.4 The second model simulations

We note that in the following simulations, P_{R3} is the real probability in the luminal universe $\mathbf{G_3}$ for $(v = c)$ in yellow in the simulations, where we have $\forall P_{R3} : 0 \leq P_{R3} \leq 1$ and that it will be included in the final most general model of *MCPP*. Thus, the current model is a simplified second model. The simulations from **Figures 1–3** illustrate the second and more general model.

2. The metarelativistic complex probability paradigm (*MCPP*): a more general third model

In this section, we will develop the third more general model of *MCPP* with all its parameters [1–42].

2.1 The real and imaginary probabilities

Here, and in this third *MCPP* model, v_1 is always the velocity of a body in R_1 with $0 \leq v_1 < c$ and is a random variable that follows any possible probability distribution: $PDF_1(\overline{v}_1, \sigma_{v1})$ where $\overline{v}_1$ is the mean or the expectation of this general probability distribution of v_1 or $PDF_1(v_1)$ and σ_{v1} is its corresponding standard deviation. And v_2 is also the velocity of a body in R_2 with $c < v_2 \leq nc$ and is a random variable that follows any possible probability distribution: $PDF_2(\overline{v}_2, \sigma_{v2})$ for a determined and fixed value of n such that $\forall n, n \in \mathbb{R}^+ : n > 1 \Leftrightarrow n \in (1, +\infty)$ and where $\overline{v}_2$ is the mean or the expectation of this general probability distribution of v_2 or $PDF_2(v_2)$ and σ_{v2} is its corresponding standard deviation. Note that, PDF_1 and PDF_2 do not have here to be similar probability distributions.

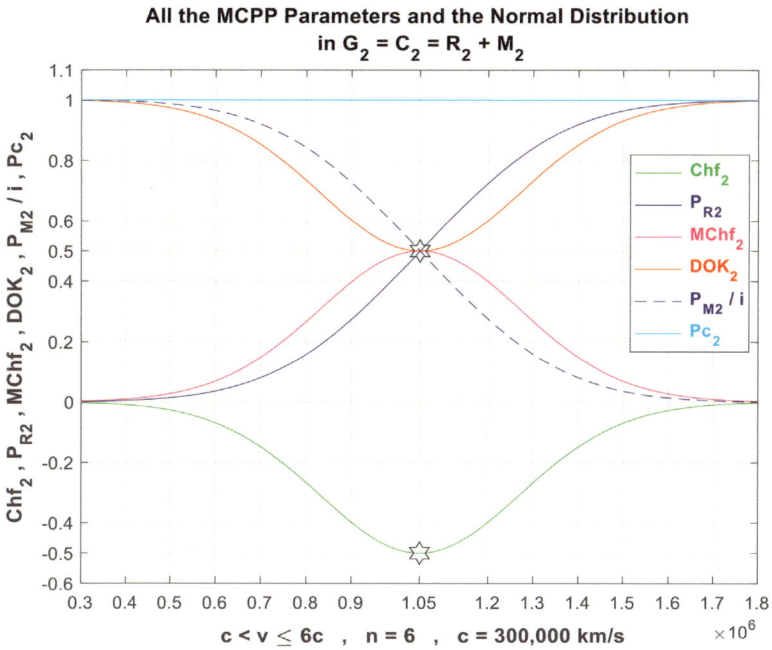

Figure 1.
The MCPP *second model parameters and the normal distribution for* $n = 6$ *in* $\mathbf{G_2}$.

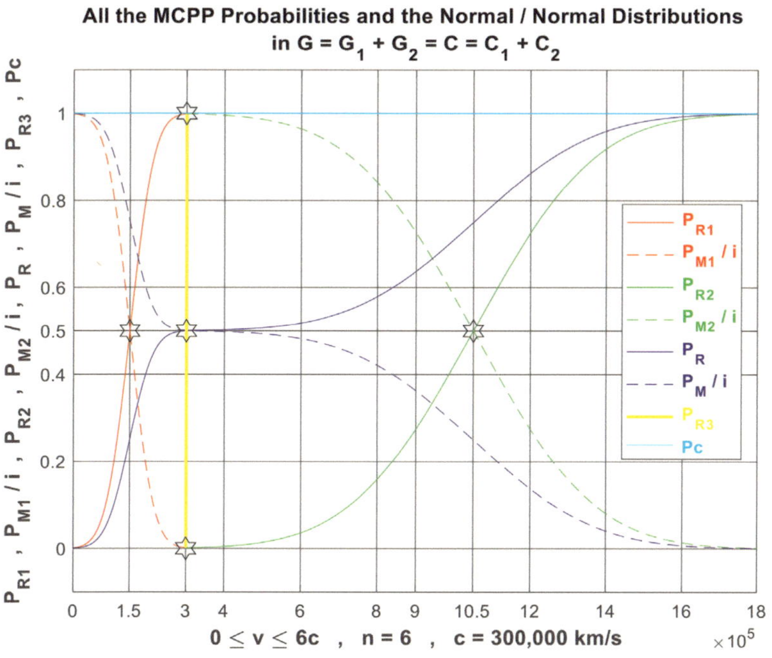

Figure 2.
The MCPP *second model probabilities and the normal/normal distributions for* $n = 6$ *in* $\mathbf{G}$.

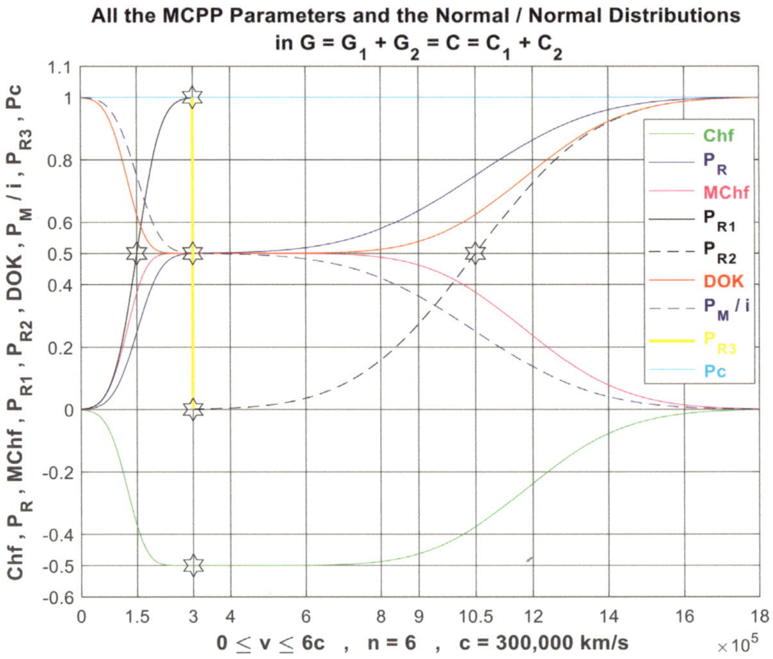

Figure 3.
The MCPP *second model parameters and the normal/normal distributions for* $n = 6$ *in* **G**.

First, we will define and calculate the real and imaginary probabilities in the universes R_1, R_2, M_1, and M_2 in the third model of $MCPP$ as follows:

$$P_{R1} = P_{rob}(0 \leq V \leq v_1) = CDF_1(0 \leq V \leq v_1) = \int_0^{v1} PDF_1(v)dv.$$

So, if $v_1 < 0 \Rightarrow P_{R1} = P_{rob}(V < 0) = CDF_1(V < 0) = 0.$

If $v_1 = 0 \Rightarrow P_{R1} = P_{rob}(V \leq 0) = CDF_1(V \leq 0) = \int_0^0 PDF_1(v)dv = 0.$

If $v_1 = \mathrm{Md}(v_1) \Rightarrow P_{R1} = P_{rob}(0 \leq V \leq \mathrm{Md}(v_1)) = CDF_1(0 \leq V \leq \mathrm{Md}(v_1)) = \int_0^{\mathrm{Md}(v_1)} PDF_1(v)dv = 0.5.$

where $\mathrm{Md}(v_1)$ is the median of the velocity v_1 probability distribution.

If $v_1 \to c^- \Rightarrow P_{R1} \to P_{rob}(0 \leq V < c) = CDF_1(0 \leq V < c) = \int_0^c PDF_1(v)dv = 1.$

If $v_1 > c \Rightarrow P_{R1} = P_{rob}(V > c) = CDF_1(V > c) = \int_0^{v1} PDF_1(v)dv = \left\{\int_0^c PDF_1(v)dv + \int_c^{v1} PDF_1(v)dv\right\} = (1 + 0) = 1.$

And we have for the second real probability:

$$P_{R2} = P_{rob}(c < V \leq v_2) = CDF_2(c < V \leq v_2) = \int_c^{v2} PDF_2(v)dv$$

So, if $v_2 < c \Rightarrow P_{R2} = P_{rob}(V < c) = CDF_2(V < c) = 0.$

If $v_2 \to c^+ \Rightarrow P_{R2} \to \int_c^c PDF_2(v)dv = 0.$

If $v_2 = \text{Md}(v_2) \Rightarrow P_{R2} = P_{rob}(c < V \leq \text{Md}(v_2)) = CDF_2(c < V \leq \text{Md}(v_2)) = \int_c^{\text{Md}(v_2)} PDF_2(v)dv = 0.5.$

where $\text{Md}(v_2)$ is the median of the velocity v_2 probability distribution.

If $v_2 = nc \Rightarrow P_{R2} = P_{rob}(c < V \leq nc) = CDF_2(c < V \leq nc) = \int_c^{nc} PDF_2(v)dv = 1.$

If $v_2 > nc$

$$\Rightarrow P_{R2} = P_{rob}(V > nc) = CDF_2(V > nc)$$
$$= \int_c^{v2} PDF_2(v)dv = \left\{ \int_c^{nc} PDF_2(v)dv + \int_{nc}^{v2} PDF_2(v)dv \right\} = (1+0) = 1$$

Moreover, the first imaginary probability is:

$$P_{M1} = i(1 - P_{R1}) = i[1 - P_{rob}(0 \leq V \leq v_1)] = i[1 - CDF_1(0 \leq V \leq v_1)] = iCDF_1(v_1 < V < c)$$
$$= i\left[1 - \int_0^{v1} PDF_1(v)dv\right] = i\int_{v1}^{c} PDF_1(v)dv$$

So, if $v_1 < 0 \Rightarrow$

$P_{M1} = i\int_{v1}^{c} PDF_1(v)dv = i\left\{\int_{v1}^{0} PDF_1(v)dv + \int_0^c PDF_1(v)dv\right\} = i(0+1) = i \Rightarrow P_{M1}/i = 1.$

If $v_1 = 0$

$\Rightarrow P_{M1} = i[1 - P_{rob}(V \leq 0)] = i[1 - CDF_1(V \leq 0)] = i(1-0) = i \Rightarrow P_{M1}/i = 1.$

If $v_1 = \text{Md}(v_1)$

$$\Rightarrow P_{M1} = i[1 - P_{rob}(0 \leq V \leq \text{Md}(v_1))] = i[1 - CDF_1(0 \leq V \leq \text{Md}(v_1))]$$
$$= i\left[1 - \int_0^{\text{Md}(v_1)} PDF_1(v)dv\right] = i\int_{\text{Md}(v_1)}^{c} PDF_1(v)dv = i(1-0.5) = 0.5i \Rightarrow P_{M1}/i = 0.5$$

If $v_1 \to c^- \Rightarrow P_{M1} \to i[1 - P_{rob}(0 \leq V < c)] = i[1 - CDF_1(0 \leq V < c)] =$

$i\left[1 - \int_0^c PDF_1(v)dv\right] = i(1-1) = 0$

$\Rightarrow P_{M1}/i \to 0$

If $v_1 > c \Rightarrow P_{M1} = 0 \Rightarrow P_{M1}/i = 0.$

And we have for the second imaginary probability:

$$P_{M2} = i(1 - P_{R2}) = i[1 - P_{rob}(c < V \leq v_2)] = i[1 - CDF_2(c < V \leq v_2)] = iCDF_2(v_2 < V \leq nc)$$
$$= i\left[1 - \int_c^{v2} PDF_2(v)dv\right] = i\int_{v2}^{nc} PDF_2(v)dv$$

So, if $v_2 < c \Rightarrow$

$$P_{M2} = i \int_{v2}^{nc} PDF_2(v)dv = i\left\{ \int_{v2}^{c} PDF_2(v)dv + \int_{c}^{nc} PDF_2(v)dv \right\} = i(0+1) = i \Rightarrow P_{M2}/i = 1.$$

If $v_2 \to c^+$

$$\begin{aligned} \Rightarrow P_{M2} \to i[1 - P_{rob}(V \le v_2)] &= iP_{rob}(c < V \le nc) = iCDF_2(c < V \le nc) = i \times 1 = i \\ &= i[1 - CDF_2(V < c)] = i(1-0) = i \\ &\Rightarrow P_{M2}/i \to 1 \end{aligned}$$

If $v_2 = \text{Md}(v_2)$

$$\begin{aligned} \Rightarrow P_{M2} &= i[1 - P_{rob}(c < V \le \text{Md}(v_2))] = i[1 - CDF_2(c < V \le \text{Md}(v_2))] \\ &= i\left[1 - \int_{c}^{\text{Md}(v_2)} PDF_2(v)dv\right] = i \times \int_{\text{Md}(v_2)}^{nc} PDF_2(v)dv = i(1-0.5) = 0.5i \Rightarrow P_{M2}/i = 0.5 \end{aligned}$$

If $v_2 = nc \Rightarrow P_{M2} = i[1 - P_{rob}(c < V \le nc)] = i[1 - CDF_2(c < V \le nc)] =$

$$i\left[1 - \int_{c}^{nc} PDF_2(v)dv\right] = i(1-1) = 0$$

$$\Rightarrow P_{M2}/i = 0$$

If $v_2 > nc \Rightarrow P_{M2} = 0 \Rightarrow P_{M2}/i = 0$.

Furthermore, we have $\mathcal{R} = R_1(0 \le v < c) + R_2(c < v \le nc),\ \forall n, n \in \mathbb{R}^+ : n > 1 \Leftrightarrow n \in (1, +\infty)$.

Now, let $P_R = \frac{P_{R1}+P_{R2}}{2}$ and it is equal to half of the sum of the cumulative probability that $0 \le V \le v_1$ in R_1 and the cumulative probability that $c < V \le v_2$ in R_2.

$$\Rightarrow P_R = \frac{CDF_1(0 \le V \le v_1) + CDF_2(c < V \le v_2)}{2} = \frac{1}{2}\left\{ \int_{0}^{v1} PDF_1(v)dv + \int_{c}^{v2} PDF_2(v)dv \right\}$$

Hence, we have in $\mathbf{G} = \mathcal{C} = \mathcal{R} + \mathcal{M} = \mathbf{G_1} + \mathbf{G_2}$: $0 \le v \le nc$ with $v \ne c$.

So, if $0 \le v < c \Rightarrow P_{R1} = P_{rob}(0 \le V \le v) = CDF_1(0 \le V \le v)$.

And $P_{R2} = P_{rob}(V \le v) = P_{rob}(V < c) = CDF_2(V < c) = 0$

$$\Rightarrow P_R = \frac{CDF_1(0 \le V \le v) + 0}{2} = \frac{CDF_1(0 \le V \le v)}{2} = \frac{P_{R1}}{2}$$

Therefore, we say here that we are working in the real probability universe $\mathcal{R} = R_1$ alone.

And if $c < v \le nc \Rightarrow P_{R1} = P_{rob}(V > c) = CDF_1(V > c) = 1$.

And $P_{R2} = P_{rob}(c < V \le v) = CDF_2(c < V \le v)$

$$\Rightarrow P_R = \frac{1 + CDF_2(c < V \le v)}{2} = \frac{1 + P_{R2}}{2}$$

Therefore, we say here that we are working in the real probability universe $\mathcal{R} = R_2$ alone.

And if $0 \le v \le nc$ with $v \ne c \Rightarrow P_{R1} = P_{rob}(0 \le V \le v) = CDF_1(0 \le V \le v)$.
And $P_{R2} = P_{rob}(c < V \le v) = CDF_2(c < V \le v)$

$$\Rightarrow P_R = \frac{CDF_1(0 \le V \le v) + CDF_2(c < V \le v)}{2} = \frac{P_{R1} + P_{R2}}{2}$$

Therefore, we say here that we are working in the real probability universe $\mathcal{R} = R_1 + R_2$.

And consequently, we can deduce from the above the real probability in the probability universe $\mathcal{R} = R_1 + R_2$ for special velocity cases as follows:

if $v < 0 \Rightarrow P_R = \frac{CDF_1(V<0)}{2} = \frac{0}{2} = 0$.

if $v = \mathrm{Md}(v_1) \Rightarrow P_R = \frac{CDF_1(0 \le V \le \mathrm{Md}(v_1)) + CDF_2(V<c)}{2} = \frac{0.5+0}{2} = 0.25$.

if $v \to c^- \Rightarrow P_R \to \frac{CDF_1(0 \le V < c) + CDF_2(V<c)}{2} = \frac{1+0}{2} = 0.5$.

if $v = \mathrm{Md}(v_2) \Rightarrow P_R = \frac{CDF_1(0 \le V < c) + CDF_2(c < V \le \mathrm{Md}(v_2))}{2} = \frac{1+0.5}{2} = 0.75$.

if $v = nc \Rightarrow P_R = \frac{CDF_1(0 \le V < c) + CDF_2(c < V \le nc)}{2} = \frac{1+1}{2} = 1$.

where $\mathrm{Md}(v_1)$ and $\mathrm{Md}(v_2)$ are the medians of the velocities probabilities distributions.

Additionally, we have $\mathcal{M} = M_1(0 \le v < c) + M_2(c < v \le nc)$, $\forall n, n \in \mathbb{R}^+ : n > 1 \Leftrightarrow n \in (1, +\infty)$.

Now, let $P_M = \frac{P_{M1}+P_{M2}}{2}$ and it is equal to half of the sum of the complement of the cumulative probability that $0 \le V \le v_1$ in M_1 and the complement of the cumulative probability that $c < V \le v_2$ in M_2.

$$\Rightarrow P_M = \frac{i(1-P_{R1}) + i(1-P_{R2})}{2} = \frac{2i - i(P_{R1}+P_{R2})}{2} = i - \frac{i(P_{R1}+P_{R2})}{2} = i\left[1 - \frac{(P_{R1}+P_{R2})}{2}\right] = i(1-P_R)$$

$$\Rightarrow P_M = \frac{i[1 - CDF_1(0 \le V \le v_1)] + i[1 - CDF_2(c < V \le v_2)]}{2}$$
$$= \frac{i}{2}\left\{\int_0^{v1}[1 - PDF_1(v)]dv + \int_c^{v2}[1 - PDF_2(v)]dv\right\} = \frac{i}{2}\left\{\int_{v1}^{c} PDF_1(v)dv + \int_{v2}^{nc} PDF_2(v)dv\right\}$$

Thus, we have in $\mathbf{G} = \mathcal{C} = \mathcal{R} + \mathcal{M} = \mathbf{G_1} + \mathbf{G_2}$: $0 \le v \le nc$ with $v \ne c$.
So, if $0 \le v < c \Rightarrow P_{M1} = i[1 - P_{rob}(0 \le V \le v)] = i[1 - CDF_1(0 \le V \le v)]$.
And $P_{M2} = i[1 - P_{rob}(0 \le V \le v)] = i[1 - CDF_2(V<c)] = i(1-0) = i$

$$\Rightarrow P_M = \frac{i[1 - CDF_1(0 \le V \le v)] + i}{2} = \frac{i + P_{M1}}{2} = i\left[1 - \frac{P_{R1}}{2}\right]$$

Therefore, we say here that we are working in the imaginary probability universe $\mathcal{M} = M_1$ alone.

And if $c < v \le nc \Rightarrow P_{M1} = i[1 - P_{rob}(V > c)] = i[1 - CDF_1(V > c)] = i(1-1) = 0$.
And $P_{M2} = i[1 - P_{rob}(c < V \le v)] = i[1 - CDF_2(c < V \le v)]$

$$\Rightarrow P_M = \frac{0 + i[1 - CDF_2(c < V \le v)]}{2} = \frac{P_{M2}}{2} = i\left[\frac{1 - P_{R2}}{2}\right]$$

Therefore, we say here that we are working in the imaginary probability universe $\mathcal{M} = M_2$ alone.

And if
$0 \le v \le nc$ with $v \ne c \Rightarrow P_{M1} = i[1 - P_{rob}(0 \le V \le v)] = i[1 - CDF_1(0 \le V \le v)]$.
And $P_{M2} = i[1 - P_{rob}(c < V \le v)] = i[1 - CDF_2(c < V \le v)] \Rightarrow$
$P_M = \frac{i[1-CDF_1(0 \le V < c)]+i[1-CDF_2(c < V \le v)]}{2} = \frac{P_{M1}+P_{M2}}{2} = i\left[1 - \frac{P_{R1}+P_{R2}}{2}\right] = i[1 - P_R]$.
Therefore, we say here that we are working in the imaginary probability universe $\mathcal{M} = M_1 + M_2$.
And consequently, we can deduce from the above the imaginary probability in the probability universe $\mathcal{M} = M_1 + M_2$ for special velocity cases as follows:
if $v < 0 \Rightarrow P_M = i\left[1 - \frac{CDF_1(V<0)}{2}\right] = i\left[1 - \frac{0}{2}\right] = i \Rightarrow P_M/i = 1$.
if $v = \mathrm{Md}(v_1) \Rightarrow P_M = i\left[1 - \frac{CDF_1(0 \le V \le \mathrm{Md}(v_1))+CDF_2(V<c)}{2}\right] = i\left[1 - \frac{0.5+0}{2}\right] = 0.75i$

$$\Rightarrow P_M/i = 0.75$$

if $v \to c^- \Rightarrow P_M \to i\left[1 - \frac{CDF_1(0 \le V < c)+CDF_2(V<c)}{2}\right] = i\left[1 - \frac{1+0}{2}\right] = 0.5i \Rightarrow P_M/i \to 0.5$.
if $v = \mathrm{Md}(v_2) \Rightarrow P_M = i\left[1 - \frac{CDF_1(0 \le V < c)+CDF_2(c < V \le \mathrm{Md}(v_2))}{2}\right] = i\left[1 - \frac{1+0.5}{2}\right] = 0.25i$

$$\Rightarrow P_M/i = 0.25$$

if $v = nc \Rightarrow P_M = i\left[1 - \frac{CDF_1(0 \le V < c)+CDF_2(c < V \le nc)}{2}\right] = i\left[1 - \frac{1+1}{2}\right] = i(1-1) = 0$

$$\Rightarrow P_M/i = 0$$

where $\mathrm{Md}(v_1)$ and $\mathrm{Md}(v_2)$ are the medians of the velocities probabilities distributions.
Therefore, for any value of $0 \le v \le nc$ with $v \ne c$, we can write without any confusion that:
$P_{M1} = i(1 - P_{R1})$ and $P_{R1} = 1 - P_{M1}/i$; hence, M_1 is the imaginary complementary probability universe to the real probability universe R_1.
And $P_{M2} = i(1 - P_{R2})$ and $P_{R2} = 1 - P_{M2}/i$; hence, M_2 is the imaginary complementary probability universe to the real probability universe R_2.
Moreover, in all cases and for any value of $v : 0 \le v \le nc$ with $v \ne c$, we have:
$P_R = \frac{P_{R1}+P_{R2}}{2}$ where $\mathcal{R} = R_1 + R_2$.
And $P_M = \frac{P_{M1}+P_{M2}}{2}$ where $\mathcal{M} = M_1 + M_2$.
We can check that:

$$P_M = \frac{i(1-P_{R1}) + i(1-P_{R2})}{2} = \frac{2i - i(P_{R1}+P_{R2})}{2} = i - \frac{i(P_{R1}+P_{R2})}{2} = i\left[1 - \frac{(P_{R1}+P_{R2})}{2}\right] = i(1-P_R)$$

Hence, $\mathcal{M}$ is the imaginary complementary probability universe to the real probability universe $\mathcal{R}$.
Moreover, we have in $\mathbf{G} = \mathcal{C} = \mathcal{R} + \mathcal{M}$ where $0 \le v \le nc$ with $v \ne c$,

$$\forall n, n \in \mathbb{R}^+ : n > 1 \Leftrightarrow n \in (1, +\infty).$$

Then, $\mathcal{C} = (R_1 + R_2) + (M_1 + M_2) = (R_1 + M_1) + (R_2 + M_2) = \mathcal{C}_1 + \mathcal{C}_2$.

In fact, in $\mathcal{C}_1$ we have: $Pc_1 = P_{R1} + P_{M1}/i = P_{R1} + (1 - P_{R1}) = 1$.
And, in $\mathcal{C}_2$ we have: $Pc_2 = P_{R2} + P_{M2}/i = P_{R2} + (1 - P_{R2}) = 1$.
And, in $\mathcal{C}$ we have:

$$\begin{aligned} Pc = P_R + P_M/i &= \frac{P_{R1} + P_{R2}}{2} + \left[\frac{P_{M1} + P_{M2}}{2}\right]/i = \frac{P_{R1} + P_{R2}}{2} + \left[\frac{i(1 - P_{R1}) + i(1 - P_{R2})}{2}\right]/i \\ &= \frac{P_{R1} + P_{R2}}{2} + \frac{(1 - P_{R1}) + (1 - P_{R2})}{2} = \frac{P_{R1} + P_{R2}}{2} + 1 - \frac{P_{R1} + P_{R2}}{2} \\ &= 1 \end{aligned}$$

We can calculate Pc using this method also:

$$\begin{aligned} Pc &= P_R + P_M/i \\ &= \frac{P_{R1} + P_{R2}}{2} + \left[\frac{P_{M1} + P_{M2}}{2}\right]/i = \frac{P_{R1} + P_{M1}/i}{2} + \frac{P_{R2} + P_{M2}/i}{2} = \frac{Pc_1}{2} + \frac{Pc_2}{2} = \frac{Pc_1 + Pc_2}{2} = \frac{1+1}{2} \\ &= 1 \end{aligned}$$

Consequently: $Pc = Pc_1 = Pc_2 = 1$, in accordance with *CPP* axioms.
Furthermore, we can state now and affirm finally that in this third model:
$\mathbf{G} = \mathbf{G_1}\ (0 \le v < c) + \mathbf{G_2}\ (c < v \le nc)$ that means that the total universe **G** is the sum of the real subluminal universe $\mathbf{G_1}$ and the imaginary superluminal universe or metauniverse $\mathbf{G_2}$.

1. The real subluminal universe $\mathbf{G_1}$ corresponds to the complex probability universe $\mathcal{C}_1$, which is also subluminal; hence, $\mathbf{G_1} = \mathcal{C}_1 = R_1 + M_1$ with $(0 \le v < c)$.

2. And the imaginary superluminal universe $\mathbf{G_2}$ or metauniverse corresponds to the complex probability universe $\mathcal{C}_2$, which is also superluminal; hence, $\mathbf{G_2} = \mathcal{C}_2 = R_2 + M_2$ with $(c < v \le nc)$.

Therefore,

$$P_{G1} = Pc_1 = P_{R1} + P_{M1}/i = P_{R1} + (1 - P_{R1}) = 1 \text{ and}$$

$$P_{G2} = Pc_2 = P_{R2} + P_{M2}/i = P_{R2} + (1 - P_{R2}) = 1.$$

Consequently, the complex total universe $\mathbf{G} = \mathbf{G_1}\ (0 \le v < c) + \mathbf{G_2}\ (c < v \le nc)$, which is the sum of the universe and the metauniverse corresponds to the complex probability universe $\mathcal{C}$ having:
$\mathbf{G} = \mathcal{C} = \mathcal{R} + \mathcal{M} = (R_1 + R_2) + (M_1 + M_2) = (R_1 + M_1) + (R_2 + M_2)$
$= \mathcal{C}_1\ (0 \le v < c) + \mathcal{C}_2\ (c < v \le nc) = \mathbf{G_1}\ (0 \le v < c) + \mathbf{G_2}\ (c < v \le nc)$ with $0 \le v \le nc$ and $v \ne c$,
Hence,

$$P_G = Pc = \frac{P_{G1} + P_{G2}}{2} = \frac{Pc_1 + Pc_2}{2} = \frac{1+1}{2} = 1$$

Consequently: $Pc = 1$, in accordance with *CPP* axioms.
Thus, we can conclude that, by adding the complementary imaginary probabilities universes M_1,M_2 and $\mathcal{M}$ to the real probabilities universes R_1,R_2 and $\mathcal{R}$ then all

random phenomena in the complex probabilities' universes $\mathcal{C}_1$, $\mathcal{C}_2$, and $\mathcal{C}$, and hence in the subluminal universe $\mathbf{G_1}$, in the superluminal universe $\mathbf{G_2}$, and in the total and complex universe $\mathbf{G}$, become absolutely and perfectly deterministic with probabilities expressed totally as follows:

$Pc = Pc_1 = Pc_2 = 1$ and $P_G = P_{G1} = P_{G2} = 1$.

2.2 The *MCPP* parameters of the third model

The *MCPP* parameters in this third model are similar to those of the first and second models and this is done by including the probabilities P_R and P_M corresponding to the third model.

2.3 The deterministic cases and the *MCPP* parameters of the third model

The deterministic cases in this third model are similar to those of the first and second models and this is done by taking into consideration the probabilities P_R and P_M pertaining and corresponding to the third model.

2.4 The third model simulations

We note that in the following simulations, P_{R3} is the real probability in the luminal universe $\mathbf{G_3}$ for $(v = c)$ in yellow in the simulations, where we have $\forall P_{R3} : 0 \leq P_{R3} \leq 1$, and that it will be included in the final most general model of *MCPP*. Thus, the current model is a simplified third model. The simulations from **Figures 4–6** illustrate the more general third model of *MCPP*.

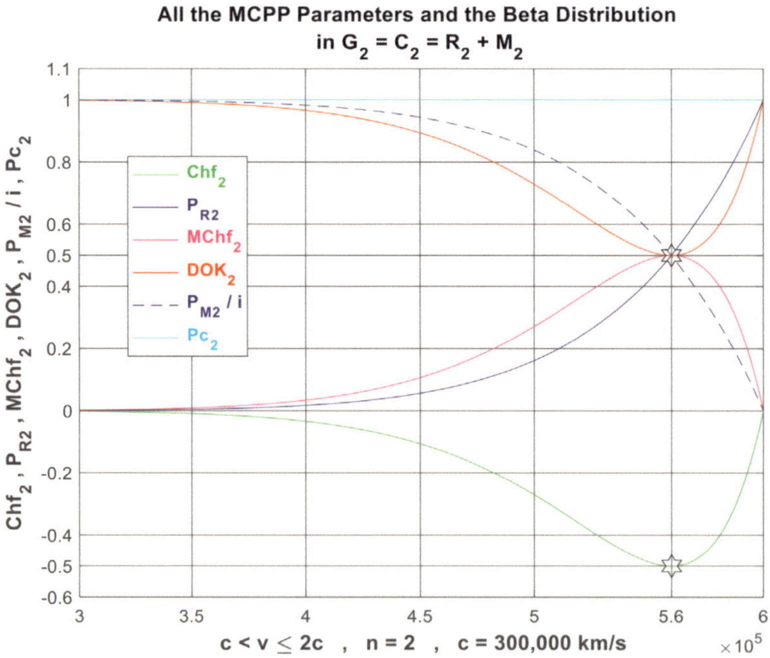

Figure 4.
The MCPP *third model parameters and the Beta distribution for $n = 2$ in* $\mathbf{G_2}$.

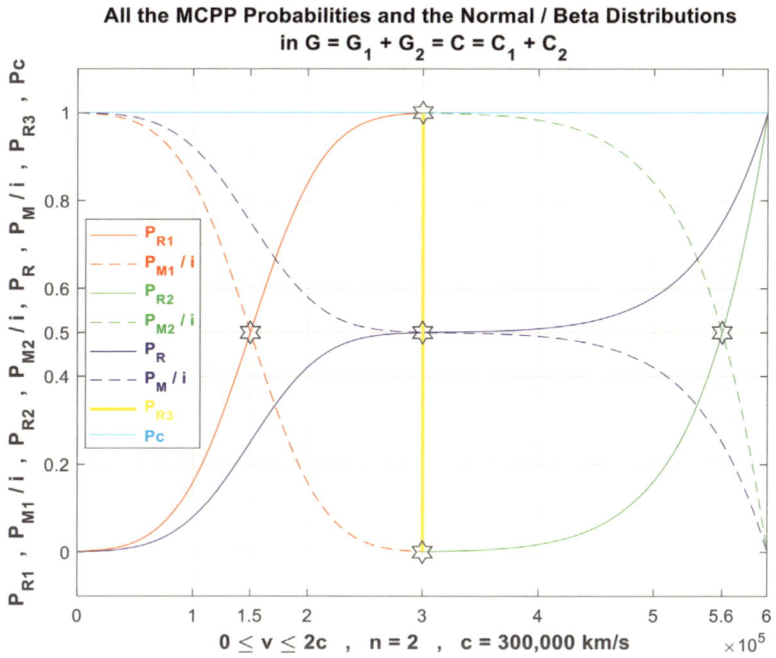

Figure 5.
The MCPP *third model probabilities and the Normal / Beta distributions for* n=2 *in G.*

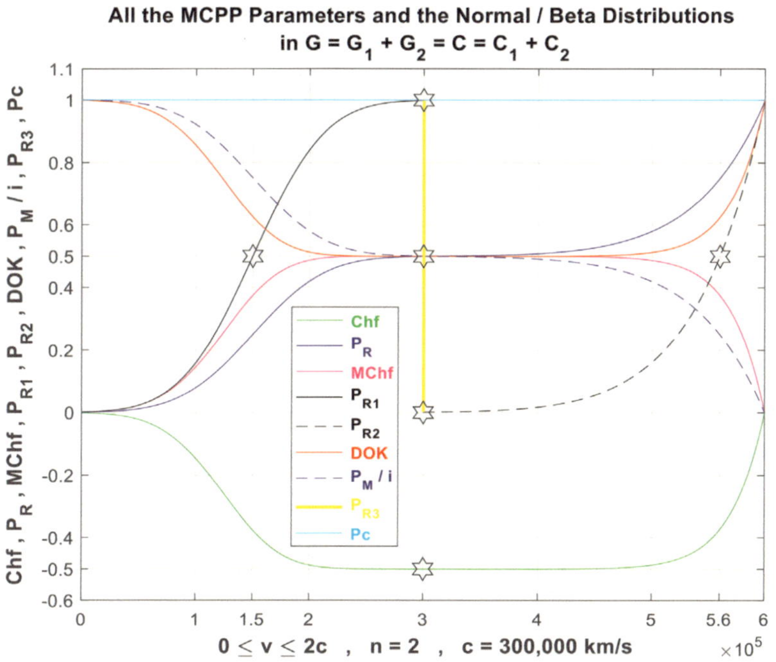

Figure 6.
The MCPP *third model parameters and the normal / Beta distributions for $n = 2$ in **G**.*

3. The final and most general model: including the case of electromagnetic waves

In this section, we will develop the final most general model of *MCPP* with all its parameters [1–42].

3.1 The real and imaginary probabilities

Here, and in this final *MCPP* model, f is the frequency of the electromagnetic waves in $\mathbf{G}_3$ since the velocity of all electromagnetic waves is always c with $(L_b = 0\ Hz) \le f \le (U_b = 10^{24} Hz)$ and is a random variable that follows any possible probability distribution: $PDF_3(\bar{f}, \sigma_f)$. Knowing that L_b is the lower bound of the frequency in the probability distribution and U_b is the upper bound of the frequency in the probability distribution. Additionally, $\bar{f}$ is the mean or the expectation of this general probability distribution of f or $PDF_3(f)$ and σ_f is its corresponding standard deviation. Moreover, the annihilation of two real particles in $\mathbf{G}_1$ or of two imaginary particles in $\mathbf{G}_2$ can lead to the creation of electromagnetic waves in $\mathbf{G} = \mathbf{G}_1 + \mathbf{G}_3 + \mathbf{G}_2$, just as proved in the theory of Metarelativity.

First, we will define and calculate the real and imaginary probabilities in the universes R_3 and M_3 in the final model of *MCPP* as follows:

$$P_{R3} = P_{rob}(v = c \text{ and } L_b \le F \le f_E) = CDF_3(v = c \text{ and } L_b \le F \le f_E) = \int_{L_b}^{f_E} PDF_3(v = c, f) df$$

where f_E is a certain value of the frequency f of the electromagnetic wave.

So, we have the real probability in R_3: if $v \ne c \Rightarrow P_{R3} = CDF_3(v \ne c) = 0$.

And if $v = c$ then:

If $f_E \le L_b \Rightarrow P_{R3} = P_{rob}(F \le L_b) = CDF_3(F \le L_b) = 0$.

If $f_E = \text{Md}(f) \Rightarrow P_{R3} = P_{rob}(L_b \le F \le \text{Md}(f)) = CDF_3(L_b \le F \le \text{Md}(f)) = \int_{L_b}^{\text{Md}(f)} PDF_3(f) df = 0.5$.

where $\text{Md}(f)$ is the median of the frequencies probability distribution.

If $f_E = U_b \Rightarrow P_{R3} = P_{rob}(L_b \le F \le U_b) = CDF_3(L_b \le F \le U_b) = \int_{L_b}^{U_b} PDF_3(f) df = 1$.

If $f_E > U_b$

$$\Rightarrow P_{R3} = P_{rob}(F > U_b) = CDF_3(F > U_b)$$
$$= \int_{L_b}^{f_E} PDF_3(f) df = \left\{ \int_{L_b}^{U_b} PDF_3(f) df + \int_{U_b}^{f_E} PDF_3(f) df \right\} = (1 + 0) = 1$$

Moreover, we have for the imaginary probability in M_3:

$$P_{M3} = i(1 - P_{R3}) = i[1 - P_{rob}(L_b \le F \le f_E)] = i[1 - CDF_3(L_b \le F \le f_E)] =$$
$$iCDF_3(f_E < F \le U_b) = i\left[1 - \int_{L_b}^{f_E} PDF_3(f) df\right] = i \int_{f_E}^{U_b} PDF_3(f) df.$$

So, if $v \neq c \Rightarrow P_{M3} = i[1 - CDF_3(v \neq c)] = i(1-0) = i \Rightarrow P_{M3}/i = 1$.

And if $v = c$ then:

If $f_E \leq L_b \Rightarrow P_{M3} = i \int_{f_E}^{U_b} PDF_3(f)df = i\left\{\int_{f_E}^{L_b} PDF_3(f)df + \int_{L_b}^{U_b} PDF_3(f)df\right\} = i(0+1) = i \Rightarrow P_{M3}/i = 1$

If $f_E = \text{Md}(f) \Rightarrow P_{M3} = i \times \int_{\text{Md}(f)}^{U_b} PDF_3(f)df = 0.5i \Rightarrow P_{M3}/i = 0.5$.

If $f_E = U_b \Rightarrow P_{M3} = i \int_{U_b}^{U_b} PDF_3(f)df = i \times 0 = 0 \Rightarrow P_{M3}/i = 0$.

If $f_E > U_b \Rightarrow P_{M3} = 0 \Rightarrow P_{M3}/i = 0$.

Therefore, for any value of f: $L_b \leq f \leq U_b$, we can write without any confusion that:

$$P_{M3} = i(1 - P_{R3}) \text{ and } P_{R3} = 1 - P_{M3}/i,$$

hence, M_3 is the imaginary complementary probability universe of frequencies to the real probability universe R_3 of frequencies.

Furthermore, we have

$$\mathcal{R} = R_1(0 \leq v < c) + R_2(c < v \leq nc) + R_3(v = c \text{ and } L_b \leq f \leq U_b),$$

$$\forall n, n \in \mathbb{R}^+ : n > 1 \Leftrightarrow n \in (1, +\infty).$$

Now, let $P_R = \frac{P_{R1}+P_{R2}+P_{R3}}{3}$ and it is equal to the sum of the cumulative probability that $0 \leq V \leq v_1$ in R_1 and the cumulative probability that $c < V \leq v_2$ in R_2 and the cumulative probability that $v_3 = c$ and $L_b \leq f \leq U_b$ in R_3, and all divided by 3.

$$\Rightarrow P_R = \frac{CDF_1(0 \leq V \leq v_1) + CDF_2(c < V \leq v_2) + CDF_3(V = v_3 = c \text{ and } L_b \leq F \leq f_E)}{3}$$

$$= \frac{1}{3}\left\{\int_0^{v1} PDF_1(v)dv + \int_c^{v2} PDF_2(v)dv + \int_{L_b}^{f_E} PDF_3(v_3 = c, f)df\right\}$$

Hence, we have in $\mathbf{G} = \mathcal{C} = \mathcal{R} + \mathcal{M} = \mathbf{G_1} + \mathbf{G_2} + \mathbf{G_3}$: $0 \leq v \leq nc$ and $L_b \leq f \leq U_b$.

So, if $0 \leq v < c \Rightarrow P_{R1} = P_{rob}(0 \leq V \leq v) = CDF_1(0 \leq V \leq v)$.

And $P_{R2} = P_{rob}(V \leq v) = P_{rob}(V < c) = CDF_2(V < c) = 0$.

And $P_{R3} = P_{rob}(V \neq c) = CDF_3(V \neq c) = 0$

$$\Rightarrow P_R = \frac{CDF_1(0 \leq V \leq v) + 0 + 0}{3} = \frac{CDF_1(0 \leq V \leq v)}{3} = \frac{P_{R1}}{3}$$

Therefore, we say here that we are working in the real probability universe $\mathcal{R} = R_1$ alone.

if $0 \leq v \leq c$ and $L_b \leq f \leq U_b \Rightarrow P_{R1} = P_{rob}(0 \leq V \leq v) = CDF_1(0 \leq V \leq v)$.

And $P_{R2} = P_{rob}(V \leq v) = P_{rob}(V < c^+) = CDF_2(V < c^+) = 0$.

And $P_{R3} = P_{rob}(V = c \text{ and } L_b \leq F \leq f_E) = CDF_3(V = c \text{ and } L_b \leq F \leq f_E)$

$$\Rightarrow P_R = \frac{CDF_1(0 \leq V \leq v) + 0 + CDF_3(V = c \text{ and } L_b \leq F \leq f_E)}{3} = \frac{P_{R1} + P_{R3}}{3}$$

Therefore, we say here that we are working in the real probability universe $\mathcal{R} = R_1 + R_3$.

And if $c < v \leq nc \Rightarrow P_{R1} = P_{rob}(V > c) = CDF_1(V > c) = 1$.
And $P_{R2} = P_{rob}(c < V \leq v) = CDF_2(c < V \leq v)$.
And $P_{R3} = P_{rob}(V \neq c) = CDF_3(V \neq c) = 0$

$$\Rightarrow P_R = \frac{1 + CDF_2(c < V \leq v) + 0}{3} = \frac{1 + P_{R2}}{3}$$

Therefore, we say here that we are working in the real probability universe $\mathcal{R} = R_2$ alone.
And if $c \leq v \leq nc$ and $L_b \leq f \leq U_b \Rightarrow P_{R1} = P_{rob}(V > c) = CDF_1(V > c) = 1$.
And $P_{R2} = P_{rob}(c < V \leq v) = CDF_2(c < V \leq v)$.
And $P_{R3} = P_{rob}\left(V = c \text{ and } L_b \leq F \leq f_E\right) = CDF_3\left(V = c \text{ and } L_b \leq F \leq f_E\right)$

$$\Rightarrow P_R = \frac{1 + CDF_2(c < V \leq v) + CDF_3\left(v = c \text{ and } L_b \leq F \leq f_E\right)}{3} = \frac{1 + P_{R2} + P_{R3}}{3}$$

Therefore, we say here that we are working in the real probability universe $\mathcal{R} = R_2 + R_3$.
And if $0 \leq v \leq nc$ and $L_b \leq f \leq U_b \Rightarrow P_{R1} = P_{rob}(0 \leq V \leq v) = CDF_1(0 \leq V \leq v)$.
And $P_{R2} = P_{rob}(c < V \leq v) = CDF_2(c < V \leq v)$.
And $P_{R3} = P_{rob}\left(V = c \text{ and } L_b \leq F \leq f_E\right) = CDF_3\left(V = c \text{ and } L_b \leq F \leq f_E\right)$

$$\Rightarrow P_R = \frac{CDF_1(0 \leq V \leq v) + CDF_2(c < V \leq v) + CDF_3\left(V = c \text{ and } L_b \leq F \leq f_E\right)}{3} = \frac{P_{R1} + P_{R2} + P_{R3}}{3}$$

Therefore, we say here that we are working in the real probability universe $\mathcal{R} = R_1 + R_2 + R_3$.

And consequently, we can deduce from the above the real probability in the probability universe $\mathcal{R} = R_1 + R_2 + R_3$ for special velocity and frequency cases as follows:

if $v < 0 \Rightarrow P_R = \frac{CDF_1(V<0)}{3} = \frac{0}{3} = 0$.
if $v = \text{Md}(v_1) \Rightarrow P_R = \frac{CDF_1(0 \leq V \leq \text{Md}(v_1)) + CDF_2(V<c) + CDF_3(V \neq c)}{3} = \frac{0.5+0+0}{3} = 0.1667$.
if $v \to c^- \Rightarrow P_R \to \frac{CDF_1(0 \leq V < c) + CDF_2(V<c) + CDF_3(V \neq c)}{3} = \frac{1+0+0}{3} = 0.3333$.
if $v = c \Rightarrow P_R = \frac{CDF_1(0 \leq V < c) + CDF_2(V < c^+) + CDF_3\left(V = c \text{ and } L_b \leq F \leq f_E\right)}{3}$

$$= \frac{1 + 0 + CDF_3\left(V = c \text{ and } L_b \leq F \leq f_E\right)}{3} = \frac{1 + CDF_3\left(V = c \text{ and } L_b \leq F \leq f_E\right)}{3}$$

So, if $f_E = L_b \Rightarrow P_R = \frac{1+0+0}{3} = \frac{1}{3} = 0.3333$.
if $f_E = \text{Md}(f) \Rightarrow P_R = \frac{1+0+0.5}{3} = \frac{1.5}{3} = 0.5$.
if $f_E = U_b \Rightarrow P_R = \frac{1+0+1}{3} = \frac{2}{3} = 0.6667$.
if $v = \text{Md}(v_2)$

$$\Rightarrow P_R = \frac{CDF_1(0 \leq V < c) + CDF_2(c < V \leq \text{Md}(v_2)) + CDF_3\left(V = c \text{ and } L_b \leq F \leq f_E\right)}{3}$$

$$= \frac{1 + 0.5 + CDF_3\left(V = c \text{ and } L_b \leq F \leq f_E\right)}{3} = \frac{1.5 + CDF_3\left(V = c \text{ and } L_b \leq F \leq f_E\right)}{3}$$

So, if $f_E = L_b \Rightarrow P_R = \frac{1+0.5+0}{3} = \frac{1.5+0}{3} = 0.5$.
if $f_E = \mathrm{Md}(f) \Rightarrow P_R = \frac{1+0.5+0.5}{3} = \frac{2}{3} = 0.6667$.
if $f_E = U_b \Rightarrow P_R = \frac{1+0.5+1}{3} = \frac{2.5}{3} = 0.8333$.
if $v = nc$

$$\Rightarrow P_R = \frac{CDF_1(0 \le V < c) + CDF_2(c < V \le nc) + CDF_3(V = c \text{ and } L_b \le F \le f_E)}{3}$$

$$= \frac{1 + 1 + CDF_3(V = c \text{ and } L_b \le F \le f_E)}{3} = \frac{2 + CDF_3(V = c \text{ and } L_b \le F \le f_E)}{3}$$

So, if $f_E = L_b \Rightarrow P_R = \frac{1+1+0}{3} = \frac{2}{3} = 0.6667$.
if $f_E = \mathrm{Md}(f) \Rightarrow P_R = \frac{1+1+0.5}{3} = \frac{2.5}{3} = 0.8333$.
if $f_E = U_b \Rightarrow P_R = \frac{1+1+1}{3} = \frac{3}{3} = 1$.
where $\mathrm{Md}(v_1)$ and $\mathrm{Md}(v_2)$ are the medians of the velocities probabilities distributions and $\mathrm{Md}(f)$ is the median of the frequencies probability distribution.

Furthermore, we have

$$\mathcal{M} = M_1(0 \le v < c) + M_2(c < v \le nc) + M_3(v = c \text{ and } L_b \le f \le U_b),$$

$$\forall n, n \in \mathbb{R}^+ : n > 1 \Leftrightarrow n \in (1, +\infty).$$

Now, let $P_M = \frac{P_{M1}+P_{M2}+P_{M3}}{3}$ and it is equal to the sum of the complement of the cumulative probability that $0 \le V \le v_1$ in M_1 and the complement of the cumulative probability that $c < V \le v_2$ in M_2 and the complement of the cumulative probability that $v_3 = c$ and $L_b \le f \le U_b$ in M_3, and all divided by 3.

$$\Rightarrow P_M = \frac{i[1 - CDF_1(0 \le V \le v_1)] + i[1 - CDF_2(c < V \le v_2)] + i[1 - CDF_3(V = v_3 = c \text{ and } L_b \le F \le f_E)]}{3}$$

$$= \frac{i}{3}\left\{\int_0^{v1}[1 - PDF_1(v)]dv + \int_c^{v2}[1 - PDF_2(v)]dv + \int_{L_b}^{f_E}[1 - PDF_3(v_3 = c, f)]df\right\}$$

$$= \frac{i}{3}\left\{\int_{v1}^{c} PDF_1(v)dv + \int_{v2}^{nc} PDF_2(v)dv + \int_{f_E}^{U_b} PDF_3(v_3 = c, f)df\right\}$$

$$\Rightarrow P_M = \frac{i(1 - P_{R1}) + i(1 - P_{R2}) + i(1 - P_{R3})}{3}$$

$$= \frac{3i - i(P_{R1} + P_{R2} + P_{R3})}{3} = i - \frac{i(P_{R1} + P_{R2} + P_{R3})}{3} = i\left[1 - \frac{(P_{R1} + P_{R2} + P_{R3})}{3}\right] = i(1 - P_R)$$

We have in $\mathbf{G} = \mathcal{C} = \mathcal{R} + \mathcal{M} = \mathbf{G_1} + \mathbf{G_2} + \mathbf{G_3}$: $0 \le v \le nc$ and $L_b \le f \le U_b$.
So, if $0 \le v < c \Rightarrow P_{M1} = i[1 - P_{rob}(0 \le V \le v)] = i[1 - CDF_1(0 \le V \le v)]$.
And $P_{M2} = i[1 - P_{rob}(0 \le V \le v)] = i[1 - CDF_2(V < c)] = i(1 - 0) = i$.
And $P_{M3} = i[1 - P_{rob}(V \ne c)] = i[1 - CDF_3(V \ne c)] = i(1 - 0) = i$

$$\Rightarrow P_M = \frac{i[1 - CDF_1(0 \le V \le v)] + i + i}{3} = \frac{2i + P_{M1}}{3} = i\left[1 - \frac{P_{R1}}{3}\right]$$

Therefore, we say here that we are working in the imaginary probability universe $\mathcal{M} = M_1$ alone.

if $0 \le v \le c$ and $L_b \le f \le U_b \Rightarrow P_{M1} = i[1 - P_{rob}(0 \le V \le v)] = i[1 - CDF_1(0 \le V \le v)]$.
And $P_{M2} = i[1 - P_{rob}(0 \le V \le v)] = i[1 - CDF_2(V < c^+)] = i(1 - 0) = i$.
And $P_{M3} = i[1 - P_{rob}(V = c \text{ and } L_b \le F \le f_E)] =$
$i[1 - CDF_3(V = c \text{ and } L_b \le F \le f_E)]$

$$\Rightarrow P_M = \frac{i[1 - CDF_1(0 \le V \le v)] + i + i[1 - CDF_3(V = c \text{ and } L_b \le F \le f_E)]}{3}$$
$$= \frac{i + (P_{M1} + P_{M3})}{3} = i\left[1 - \frac{P_{R1} + P_{R3}}{3}\right]$$

Therefore, we say here that we are working in the imaginary probability universe $\mathcal{M} = M_1 + M_3$.
And if $c < v \le nc \Rightarrow P_{M1} = i[1 - P_{rob}(V > c)] = i[1 - CDF_1(V > c)] = i(1 - 1) = 0$
And $P_{M2} = i[1 - P_{rob}(c < V \le v)] = i[1 - CDF_2(c < V \le v)]$.
And $P_{M3} = i[1 - P_{rob}(V \ne c)] = i[1 - CDF_3(V \ne c)] = i(1 - 0) = i$

$$\Rightarrow P_M = \frac{0 + i[1 - CDF_2(c < V \le v)] + i}{3} = \frac{i + P_{M2}}{3} = i\left[\frac{2 - P_{R2}}{3}\right]$$

Therefore, we say here that we are working in the imaginary probability universe $\mathcal{M} = M_2$ alone.
And if $c \le v \le nc$ and $L_b \le f \le U_b$

$$\Rightarrow P_{M1} = i[1 - P_{rob}(V > c)] = i[1 - CDF_1(V > c)] = i(1 - 1) = 0$$

And $P_{M2} = i[1 - P_{rob}(c < V \le v)] = i[1 - CDF_2(c < V \le v)]$.
And $P_{M3} = i[1 - P_{rob}(V = c \text{ and } L_b \le F \le f_E)] = i[1 - CDF_3(V = c \text{ and } L_b \le F \le f_E)]$

$$\Rightarrow P_M = \frac{0 + i[1 - CDF_2(c < V \le v)] + i[1 - CDF_3(V = c \text{ and } L_b \le F \le f_E)]}{3} = \frac{P_{M2} + P_{M3}}{3}$$
$$= i\left[\frac{2 - (P_{R2} + P_{R3})}{3}\right]$$

Therefore, we say here that we are working in the imaginary probability universe $\mathcal{M} = M_2 + M_3$.
And if $0 \le v \le nc$ and $L_b \le f \le U_b$
$\Rightarrow P_{M1} = i[1 - P_{rob}(0 \le V \le v)] = i[1 - CDF_1(0 \le V \le v)]$.
And $P_{M2} = i[1 - P_{rob}(c < V \le v)] = i[1 - CDF_2(c < V \le v)]$.
And $P_{M3} = i[1 - P_{rob}(V = c \text{ and } L_b \le F \le f_E)] =$
$i[1 - CDF_3(V = c \text{ and } L_b \le F \le f_E)]$

$$\Rightarrow P_M = \frac{i[1 - CDF_1(0 \le V < c)] + i[1 - CDF_2(c < V \le v)] + i[1 - CDF_3(V = c \text{ and } L_b \le F \le f_E)]}{3}$$
$$= \frac{P_{M1} + P_{M2} + P_{M3}}{3} = i\left[1 - \frac{P_{R1} + P_{R2} + P_{R3}}{3}\right] = i[1 - P_R]$$

Therefore, we say here that we are working in the imaginary probability universe $\mathcal{M} = M_1 + M_2 + M_3$.

And consequently, we can deduce from the above the imaginary probability in the probability universe $\mathcal{M} = M_1 + M_2 + M_3$ for special velocity and frequency cases as follows:

if $v < 0 \Rightarrow P_M = i\left[1 - \frac{CDF_1(V<0)}{3}\right] = i\left[1 - \frac{0}{3}\right] = i \Rightarrow P_M/i = 1.$

if $v = \text{Md}(v_1) \Rightarrow P_M = i\left[1 - \frac{CDF_1(0 \le V \le \text{Md}(v_1)) + CDF_2(V<c) + CDF_3(V \ne c)}{3}\right] = i\left[1 - \frac{0.5+0+0}{3}\right] = 0.8333i \Rightarrow P_M/i = 0.8333.$

if $v \to c^- \Rightarrow P_M \to i\left[1 - \frac{CDF_1(0 \le V < c) + CDF_2(V<c) + CDF_3(V \ne c)}{3}\right] = i\left[1 - \frac{1+0+0}{3}\right] = \frac{2i}{3} = 0.6667i \Rightarrow P_M/i \to 0.6667.$

if $v = c$

$$\Rightarrow P_M = i\left[1 - \frac{CDF_1(0 \le V < c) + CDF_2(V < c^+) + CDF_3(V = c \text{ and } L_b \le F \le f_E)}{3}\right]$$

$$= i\left[1 - \frac{1 + 0 + CDF_3(V = c \text{ and } L_b \le F \le f_E)}{3}\right] = i\left[\frac{2 - CDF_3(V = c \text{ and } L_b \le F \le f_E)}{3}\right]$$

So, if $f_E = L_b \Rightarrow P_M = i\left[\frac{2-0}{3}\right] = 0.6667i \Rightarrow P_M/i = 0.6667.$

if $f_E = \text{Md}(f) \Rightarrow P_M = i\left[\frac{2-0.5}{3}\right] = 0.5i \Rightarrow P_M/i = 0.5.$

if $f_E = U_b \Rightarrow P_M = i\left[\frac{2-1}{3}\right] = 0.3333i \Rightarrow P_M/i = 0.3333.$

If $v = \text{Md}(v_2) \Rightarrow P_M = i\left[1 - \frac{CDF_1(0 \le V < c) + CDF_2(c < V \le \text{Md}(v_2)) + CDF_3(V = c \text{ and } L_b \le F \le f_E)}{3}\right]$

$$= i\left[1 - \frac{1 + 0.5 + CDF_3(V = c \text{ and } L_b \le F \le f_E)}{3}\right] = i\left[0.5 - \frac{CDF_3(V = c \text{ and } L_b \le F \le f_E)}{3}\right].$$

So, if $f_E = L_b \Rightarrow P_M = i\left[0.5 - \frac{0}{3}\right] = 0.5i \Rightarrow P_M/i = 0.5.$

if $f_E = \text{Md}(f) \Rightarrow P_M = i\left[0.5 - \frac{0.5}{3}\right] = 0.3333i \Rightarrow P_M/i = 0.3333.$

if $f_E = U_b \Rightarrow P_M = i\left[0.5 - \frac{1}{3}\right] = 0.1667i \Rightarrow P_M/i = 0.1667.$

if $v = nc$

$$\Rightarrow P_M = i\left[1 - \frac{CDF_1(0 \le V < c) + CDF_2(c < V \le nc) + CDF_3(V = c \text{ and } L_b \le F \le f_E)}{3}\right]$$

$$= i\left[1 - \frac{1 + 1 + CDF_3(V = c \text{ and } L_b \le F \le f_E)}{3}\right] = i\left[\frac{1 - CDF_3(V = c \text{ and } L_b \le F \le f_E)}{3}\right]$$

So, if $f_E = L_b \Rightarrow P_M = i\left[\frac{1-0}{3}\right] = \frac{i}{3} = 0.3333i \Rightarrow P_M/i = 0.3333.$

if $f_E = \text{Md}(f) \Rightarrow P_M = i\left[\frac{1-0.5}{3}\right] = 0.1667i \Rightarrow P_M/i = 0.1667.$

if $f_E = U_b \Rightarrow P_M = i\left[\frac{1-1}{3}\right] = 0 \Rightarrow P_M/i = 0.$

where $\text{Md}(v_1)$ and $\text{Md}(v_2)$ are the medians of the velocities probabilities distributions and $\text{Md}(f)$ is the median of the frequencies probability distribution.

Therefore, for any value of $0 \le v \le nc$, we can write without any confusion that:

$P_{M1} = i(1 - P_{R1})$ and $P_{R1} = 1 - P_{M1}/i$; hence, M_1 is the imaginary complementary probability universe to the real probability universe R_1.

And $P_{M2} = i(1 - P_{R2})$ and $P_{R2} = 1 - P_{M2}/i$; hence, M_2 is the imaginary complementary probability universe to the real probability universe R_2.

And $P_{M3} = i(1 - P_{R3})$ and $P_{R3} = 1 - P_{M3}/i$; hence, M_3 is the imaginary complementary probability universe to the real probability universe R_3.

Moreover, in all cases and for any value of $v : 0 \le v \le nc$, we have:

$P_R = \frac{P_{R1}+P_{R2}+P_{R3}}{3}$ where $\mathcal{R} = R_1 + R_2 + R_3$.

And $P_M = \frac{P_{M1}+P_{M2}+P_{M3}}{3}$ where $\mathcal{M} = M_1 + M_2 + M_3$.

Thus, we can check that:

$$P_M = \frac{i(1-P_{R1}) + i(1-P_{R2}) + i(1-P_{R3})}{3}$$
$$= \frac{3i - i(P_{R1}+P_{R2}+P_{R3})}{3} = i - \frac{i(P_{R1}+P_{R2}+P_{R3})}{3} = i\left[1 - \frac{(P_{R1}+P_{R2}+P_{R3})}{3}\right] = i(1-P_R)$$

Hence, $\mathcal{M}$ is the imaginary complementary probability universe to the real probability universe $\mathcal{R}$.

Furthermore, we can state now and affirm finally that in this final *MCPP* model:

$\mathbf{G} = \mathbf{G_1}\ (0 \le v < c) + \mathbf{G_2}\ (c < v \le nc) + \mathbf{G_3}\ (v = c)$ that means that the total universe **G** is the sum of the real subluminal universe $\mathbf{G_1}$ and the imaginary superluminal universe or metauniverse $\mathbf{G_2}$ in addition to the electromagnetic waves' universe $\mathbf{G_3}$, which stands between $\mathbf{G_1}$ and $\mathbf{G_2}$.

1. The real subluminal universe $\mathbf{G_1}$ corresponds to the complex probability universe $\mathcal{C}_1$, which is also subluminal; hence, $\mathbf{G_1} = \mathcal{C}_1 = R_1 + M_1$ with $(0 \le v < c)$.

2. The imaginary superluminal universe $\mathbf{G_2}$ or metauniverse corresponds to the complex probability universe $\mathcal{C}_2$, which is also superluminal; hence, $\mathbf{G_2} = \mathcal{C}_2 = R_2 + M_2$ with $(c < v \le nc)$, $\forall n, n \in \mathbb{R}^+ : n > 1 \Leftrightarrow n \in (1, +\infty)$.

3. In addition, the luminal universe $\mathbf{G_3}$ of electromagnetic waves corresponds to the complex probability universe $\mathcal{C}_3$ of frequencies, which is also luminal; hence, $\mathbf{G_3} = \mathcal{C}_3 = R_3 + M_3$ with ($v = c$ and $L_b \le f \le U_b$).

Therefore,

$$P_{G1} = Pc_1 = P_{R1} + P_{M1}/i = P_{R1} + (1 - P_{R1}) = 1 \text{ and}$$
$$P_{G2} = Pc_2 = P_{R2} + P_{M2}/i = P_{R2} + (1 - P_{R2}) = 1 \text{ and}$$
$$P_{G3} = Pc_3 = P_{R3} + P_{M3}/i = P_{R3} + (1 - P_{R3}) = 1.$$

Consequently, the complex total universe

$$\mathbf{G} = \mathbf{G_1}(0 \le v < c) + \mathbf{G_2}(0 < v \le nc) + \mathbf{G_3}(v = c),$$

which is the sum of the universe and the metauniverse and the luminal electromagnetic waves universe and which corresponds to the complex probability universe $\mathcal{C}$ having:

$\mathbf{G} = \mathcal{C} = \mathcal{R} + \mathcal{M} = (R_1 + R_2 + R_3) + (M_1 + M_2 + M_3) = (R_1 + M_1) + (R_2 + M_2) + (R_3 + M_3)$

$= \mathcal{C}_1\ (0 \le v < c) + \mathcal{C}_2\ (c < v \le nc) + \mathcal{C}_3\ (v = c \text{ and } L_b \le f \le U_b)$

$= \mathbf{G_1}\ (0 \le v < c) + \mathbf{G_2}\ (c < v \le nc) + \mathbf{G_3}\ (v = c \text{ and } L_b \le f \le U_b)$.

Hence,

$$P_G = Pc = \frac{P_{G1} + P_{G2} + P_{G3}}{3} = \frac{Pc_1 + Pc_2 + Pc_3}{3} = \frac{1+1+1}{3} = 1$$

Additionally,

$$P_G = P_R + P_M/i = P_R + (1 - P_R) = 1$$

Consequently, $Pc = Pc_1 = Pc_2 = Pc_3 = 1$ in accordance with *CPP* axioms.

Thus, we can conclude that, by adding the complementary imaginary probabilities universes M_1,M_2,M_3, and $\mathcal{M}$ to the real probabilities' universes R_1, R_2, R_3 and $\mathcal{R}$ then all random phenomena in the complex probabilities' universes $\mathcal{C}_1$, $\mathcal{C}_2$, $\mathcal{C}_3$, and $\mathcal{C}$, and hence in the subluminal universe $\mathbf{G_1}$, in the superluminal universe $\mathbf{G_2}$, in the luminal universe $\mathbf{G_3}$ of electromagnetic waves, and in the total and complex universe **G**, become absolutely and perfectly deterministic with probabilities expressed totally as follows:

$Pc = Pc_1 = Pc_2 = Pc_3 = 1$ and $P_G = P_{G1} = P_{G2} = P_{G3} = 1$.

3.2 The *MCPP* parameters in $\mathbf{G_3} = \mathcal{C}_3 = R_3 + M_3$

In this section, we will determine and calculate all the *MCPP* parameters in $\mathbf{G_3}$ as follows:

The real probabilities in $\mathcal{R} = R_3$: $P_{R3} = P_{rob}\left(v = c \text{ and } L_b \le F \le f_E\right) =$

$CDF_3\left(v = c \text{ and } L_b \le F \le f_E\right) = \int_{L_b}^{f_E} PDF_3(v = c, f)df$ for any value of f: $L_b \le f \le U_b$.

The imaginary complementary probabilities in $\mathcal{M} = M_3$:

$$P_{M3} = i(1 - P_{R3})$$

The real complementary probabilities in $\mathcal{R} = R_3$:

$$P_{M3}/i = 1 - P_{R3}$$

The complex random vector:

$$Z_3 = P_{R3} + P_{M3}$$

The degree of our knowledge:

$$DOK_3 = |Z_3|^2 = |P_{R3} + P_{M3}|^2 = P_{R3}^2 + [P_{M3}/i]^2 = P_{R3}^2 + [1 - P_{R3}]^2$$

The chaotic factor:

$$Chf_3 = 2iP_{R3}P_{M3} = 2iP_{R3}i(1 - P_{R3}) = 2i^2P_{R3}(1 - P_{R3}) = -2P_{R3}(1 - P_{R3})$$

The magnitude of the chaotic factor:

$$MChf_3 = |Chf_3| = -2iP_{R3}P_{M3} = -2iP_{R3}i(1 - P_{R3}) = -2i^2P_{R3}(1 - P_{R3}) = 2P_{R3}(1 - P_{R3})$$

The deterministic probability in $\mathbf{G_3} = \mathcal{C}_3 = R_3 + M_3$:

$$\begin{aligned} Pc_3^2 &= [P_{R3} + P_{M3}/i]^2 = [P_{R3} + (1 - P_{R3})]^2 = 1^2 = 1 \\ &= DOK_3 - Chf_3 = 1 \\ &= DOK_3 + MChf_3 = 1 \\ &= Pc_3 \end{aligned}$$

3.3 The final *MCPP* parameters in $\mathbf{G} = \mathbf{G}_1 + \mathbf{G}_2 + \mathbf{G}_3 = \mathcal{C} = \mathcal{C}_1 + \mathcal{C}_2 + \mathcal{C}_3 = \mathcal{R} + \mathcal{M}$ of the final and most general model

The *MCPP* parameters in this final and most general model are similar to those of the first and second and third models and this is done by including the probabilities P_R and P_M corresponding to the final model. These paradigm parameters are determined and computed as follows:

The real probabilities in $\mathcal{R} = R_1 + R_2 + R_3$: $P_R = \frac{P_{R1}+P_{R2}+P_{R3}}{3}$.

The imaginary complementary probabilities in $\mathcal{M} = M_1 + M_2 + M_3$:

$$P_M = \frac{P_{M1} + P_{M2} + P_{M3}}{3} = i(1 - P_R) = i\left[1 - \frac{(P_{R1} + P_{R2} + P_{R3})}{3}\right]$$

The real complementary probabilities in $\mathcal{R} = R_1 + R_2 + R_3$:

$$P_M/i = \left[\frac{P_{M1} + P_{M2} + P_{M3}}{3}\right]/i = \frac{P_{M1}/i + P_{M2}/i + P_{M3}/i}{3} = i(1 - P_R)/i = 1 - \frac{(P_{R1} + P_{R2} + P_{R3})}{3}$$

The complex random vectors:

$$\begin{aligned} Z &= \frac{Z_1 + Z_2 + Z_3}{3} = \frac{(P_{R1} + P_{M1}) + (P_{R2} + P_{M2}) + (P_{R3} + P_{M3})}{3} = \frac{P_{R1} + P_{R2} + P_{R3}}{3} + \frac{P_{M1} + P_{M2} + P_{M3}}{3} \\ &= P_R + P_M \end{aligned}$$

The degree of our knowledge:

$$DOK = |Z|^2 = |P_R + P_M|^2 = P_R^2 + [P_M/i]^2 = P_R^2 + [1 - P_R]^2$$

The chaotic factor:

$$Chf = 2iP_R P_M = 2iP_R i(1 - P_R) = 2i^2 P_R(1 - P_R) = -2P_R(1 - P_R)$$

The magnitude of the chaotic factor:

$$MChf = |Chf| = -2iP_R P_M = -2iP_R i(1 - P_R) = -2i^2 P_R(1 - P_R) = 2P_R(1 - P_R)$$

The deterministic probability in $\mathbf{G} = \mathcal{C}$

$$\begin{aligned} Pc^2 &= [P_R + P_M/i]^2 = [P_R + (1 - P_R)]^2 = 1^2 = 1 \\ &= DOK - Chf = 1 \\ &= DOK + MChf = 1 \\ &= Pc \end{aligned}$$

3.4 The deterministic cases and the *MCPP* parameters of the final model

The deterministic cases in this final model are similar to those of the first and second and third models and this is done by taking into consideration the probabilities P_R and P_M pertaining to and corresponding to the final model.

3.5 The simulations in the universe $\mathbf{G_3} = \mathcal{C}_3 = \boldsymbol{R_3} + \boldsymbol{M_3}$

All the simulations from **Figures 7–10** illustrate *MCPP* in the luminal universe $\mathbf{G_3}$ of electromagnetic waves.

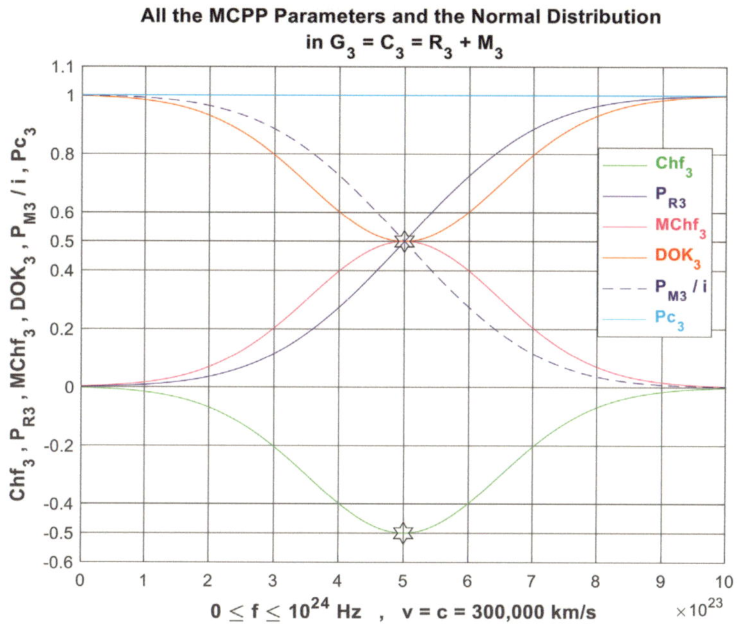

Figure 7.
The MCPP *final model parameters and the normal distribution in* $\boldsymbol{G_3}$.

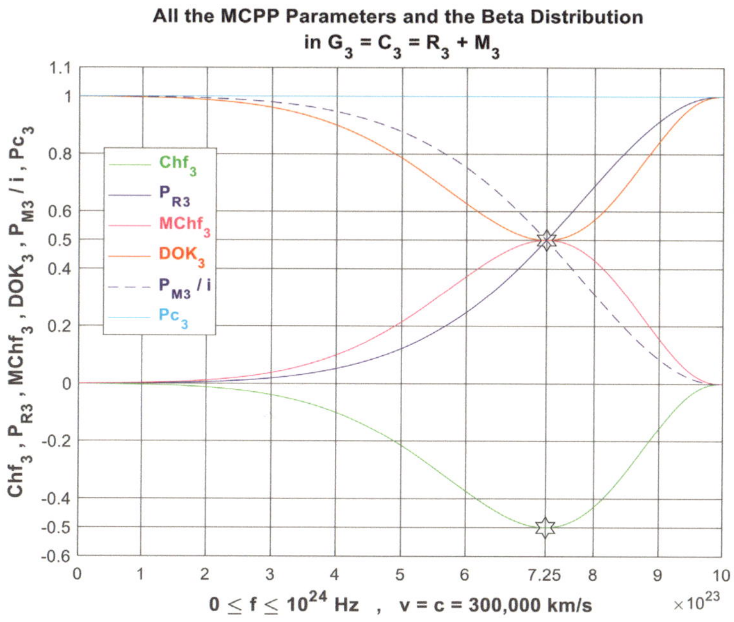

Figure 8.
The MCPP *final model parameters and the Beta distribution in* $\boldsymbol{G_3}$.

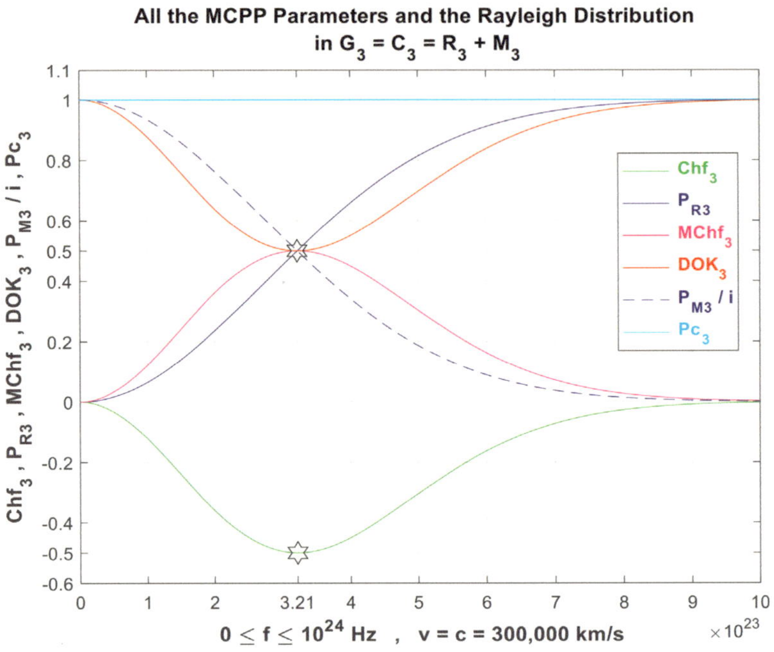

Figure 9.
The MCPP *final model parameters and the Rayleigh distribution in* G_3.

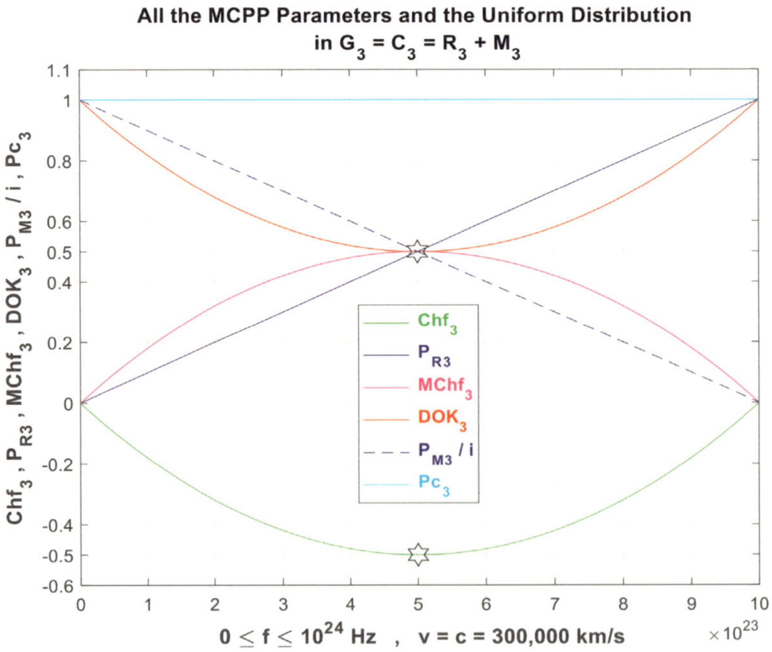

Figure 10.
The MCPP *final model parameters and the uniform distribution in* G_3.

4. Some very important consequences and advantages of the *MCPP* paradigm

In this section, we will examine, determine, and deduce some very important consequences and advantages of *MCPP* that have been developed in the previous sections of the two chapters.

4.1 Dark matter and dark energy

According to Astronomical Observations [40–42]:

The total mass-energy of the universe contains 5% ordinary matter and energy, 27% dark matter, and 68% of a form of energy known as dark energy. Thus, dark matter constitutes 85% of total mass, while dark energy plus dark matter constitutes 95% of total mass-energy content.

⇒ Dark Matter = Metamatter in $\mathbf{G_2}$.

⇒ Dark Energy = Metaenergy in $\mathbf{G_2}$.

$\mathbf{G} = \mathbf{G_1} + \mathbf{EW} + \mathbf{G_2}$

$= \mathbf{3/3} = \mathbf{1} = \mathbf{100\%}$

$\cong \mathbf{2/3}\ \{\pm i|E_{G2}| = \mathbf{1/3}\ [+i|E_{G2}|] + \mathbf{1/3}\ [-i|E_{G2}|]\} =$ **66.66% dark energy in** $\mathbf{G_2}$

$+ \cong \mathbf{1/3\ (1/30 + 9/30 = 10/30) = 33.33\%}$.

Iceberg similarity:

$\cong$ 10% of 1/3 = 1/30 = 3.33% ordinary matter and energy in $\mathbf{G_1}$.

$\cong$ 90% of 1/3 = 9/30 = 30% dark matter in $\mathbf{G_2}$.

Then $\cong$ **2/3 + 9/30 = 87/90 = 96.6667% of the total mass-energy content in G = (dark energy + dark matter) in** $\mathbf{G_2}$ (refer to **Figure 11**).

4.2 Solution of the cause-effect paradox

As it is known, we have always in the subluminal universe $\mathbf{G_1}$ (our ordinary "real" universe) the effect of any action following the cause of this action. In fact, physicists noted that if tachyons travel in the past then we will have the action cause following

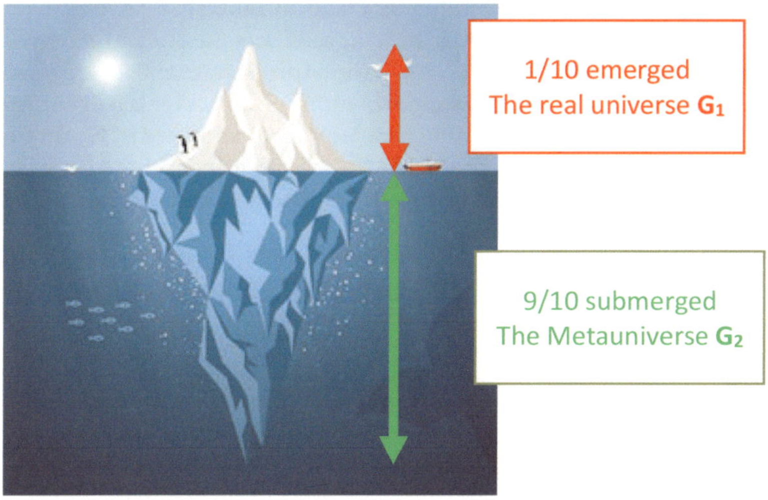

Figure 11.
The total universe $G = G_1 + EW + G_2$ *similar to an iceberg.*

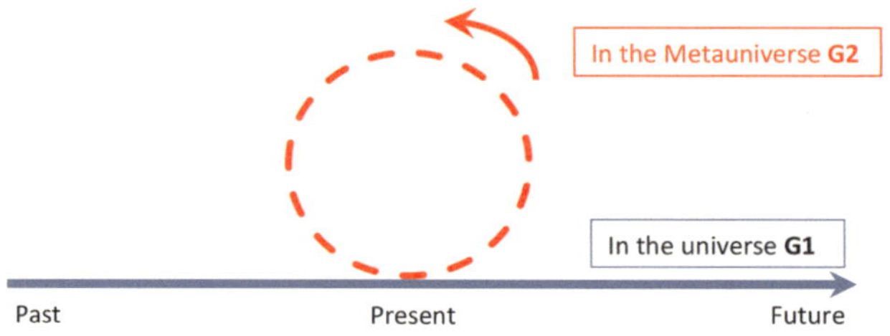

Figure 12.
The flow of time in both $\mathbf{G_1}$ and $\mathbf{G_2}$.

the action effect, which is certainly absurd. In fact, the solution of this riddle and paradox follows directly from Metarelativity, where we have $t \le 0$ is one possible solution in $\mathbf{G_2}$, then metaparticles or tachyons travel in the past in the "imaginary" superluminal metauniverse $\mathbf{G_2}$ where $t \le 0$ relative to the "real" subluminal universe $\mathbf{G_1}$, where $t \ge 0$ (**Figure 12**). This means the following:

1. No action effect before action cause in the universe $\mathbf{G_1}$ since faster-than-light particles travel in the metauniverse $\mathbf{G_2}$ and not in the universe $\mathbf{G_1}$

2. Instantaneous effect on the present in the universe $\mathbf{G_1}$ since metaparticles travel in the metauniverse $\mathbf{G_2}$ and in the relative past of the universe $\mathbf{G_1}$; hence, the instantaneous effect of the past in $\mathbf{G_2}$ on the present in $\mathbf{G_1}$ as it is shown in the figure above since the past has already occurred and its consequences and results are direct and immediate on the present and current instant.

4.3 Solution of Einstein-Podolsky-Rosen (*EPR*) paradox or of quantum entanglement

As a consequence of the previous Section 4.2 is the solution of the very famous *EPR* paradox that states if quantum entanglement were true than there should be faster-than-light particles, which are forbidden by classical Einstein's relativity. Metarelativity solves also this paradox. In fact, as it was mentioned, tachyons that travel in the past of $\mathbf{G_2}$ relatively to $\mathbf{G_1}$ have therefore instantaneous effect on $\mathbf{G_1}$. This consequence of *MCPP* explains and supports totally Alain Aspect's experiment and results on the instantaneous quantum entanglement of particles in $\mathbf{G_1}$ through the interchange of metaparticles or tachyons in $\mathbf{G_2}$, whose effect on the present is instantaneous in $\mathbf{G_1}$ since they travel relatively to $\mathbf{G_1}$ in the past in $\mathbf{G_2}$ (since $t \le 0$ is one possible solution in $\mathbf{G_2}$) (**Figure 13**).

4.4 Gravitational effect of $\mathbf{G_2}$ on $\mathbf{G_1}$ in G

As it was proved in metarelativity [1], light is the limit and the constant velocity in both $\mathbf{G_1}$ and $\mathbf{G_2}$ (refer to Section 4.7 in Chapter 1). Consequently, gravitation behaves relatively to matter in $\mathbf{G_1}$ just like to metamatter in $\mathbf{G_2}$ since it has the speed of light. Therefore, metamatter exerts gravitational effects on the matter just like ordinary matter as a result of this fact. It is the effect of metamatter on matter that we observe inside and outside galaxies. Consequently, the dark matter, which is metamatter can attract ordinary matter. This is what we are actually observing in astronomy.

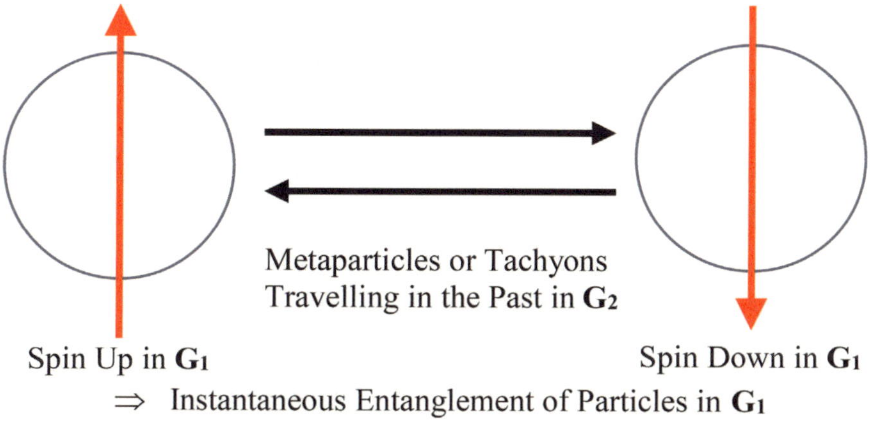

Figure 13.
Quantum entanglement of particles in the real universe G_1.

Therefore,

1. ⇒ Gravitational waves travel with the velocity of light.

2. ⇒ Light is constant and is the limit velocity in both $\mathbf{G_1}$ and $\mathbf{G_2}$.

⇒ The gravitational effect of metamatter in $\mathbf{G_2}$ on ordinary matter in $\mathbf{G_1}$.

4.5 Beneath the geometric point and beyond the infinity of time

In the Metarelativistic transformations, velocity becomes larger than the velocity of light and new Metarelativisitc equations are used to express the behavior of matter (or metamatter) inside it. In fact, starting from zero, when velocity increases, time starts to dilate as it is shown in the equation $T' = T/\sqrt{1 - \frac{v^2}{c^2}}$ and space to contract as it is shown in the equation $L' = L\sqrt{1 - \frac{v^2}{c^2}}$ and according to the well-known Einstein-Lorentz mathematical equations. If $v \to c$ then $L' \to 0$ that means when reaching the velocity of light, length at the end of contraction reaches its limits and becomes equal to zero. When velocity surpasses the barrier of light, space can start expanding as it is shown in the equation: $L' = i \times L\sqrt{\frac{v^2}{c^2} - 1}$, or it can start contracting again as it is shown in the equation: $L' = -i \times L\sqrt{\frac{v^2}{c^2} - 1}$, and this after it has reached the dimensions zero which are the dimensions of a geometric point. As a matter of fact, Euclid defined in his ELEMENTS [43] the geometric point as a geometrical entity of dimensions zero. What is smaller than zero in algebra are negative numbers. What is smaller than the geometric zero is new to us. In fact, particles in the atomic world have dimensions and the smallest particles to our knowledge are the quarks, which are the constituents of protons and neutrons. Even strings, the smallest postulated entities in String Theory, have dimensions greater than zero. Surely the dimensions of the quarks are smaller than the dimensions of protons and neutrons but they still have dimensions how small as they can be, but never the dimensions of a geometric point because zero is nothing in physics and it could not contain neither matter nor energy, except photons: photons move at the velocity of light and could never have dimensions because nature forbids that a moving body having the velocity of light has any length. The last fact was shown by Einstein in the theory of special relativity. Therefore, we can say that when space reaches zero dimensions (when $v = c$) it has reached the end of the shrinking process and what is smaller than zero in the four real dimensions of the universe $\mathbf{G_1}$ is the zero in the four imaginary dimensions of the metauniverse $\mathbf{G_2}$. Hence, after reaching zero in $\mathbf{G_1}$, space starts here in the metauniverse from zero to expand opening the field to new four imaginary dimensions as it is shown in the equation derived from the theory of Metarelativity, which is: $L' = i \times L\sqrt{\frac{v^2}{c^2} - 1}$, or it can start to contract again as it is shown in the equation: $L' = -i \times L\sqrt{\frac{v^2}{c^2} - 1}$.

Moreover, starting from zero, if speed continues to increase, time will continue to dilate in special relativity as it is shown in the equation $T' = T/\sqrt{1 - \frac{v^2}{c^2}}$. If $v \to c$ then $T' \to +\infty$ that means that if velocity reaches the velocity of light, therefore time reaches infinity. In fact, infinity in mathematics is the greatest "number" that we can ever reach while counting but we can never reach. To be more accurate, it is a symbol more than a number, since no computer could reach infinity. Infinity is not finite by nature. Infinity is extensively used in mathematics like in series and in sequences in calculus. This is why light is said to be the limit velocity and the barrier between the two geometries: the universe $\mathbf{G_1}$ and the metauniverse $\mathbf{G_2}$. In fact, if $v \to c$ then $T' \to +\infty$ and if the velocity surpasses the velocity of light: $v > c$, then time has to surpass infinity in the "real" subluminal universe $\mathbf{G_1}$: $T' > +\infty$ and hence we start counting time anew but now in the new "imaginary" superluminal metauniverse $\mathbf{G_2}$. The counting is done now using clocks set up in the metauniverse or in the four imaginary dimensions that we have already discovered in the previous metarelativistic transformations. We precise again that the new dimensions are imaginary in the sense that they contain the imaginary number i. The time measurement can start now counterclockwise because it is negative and time is said to be dilating again since $T' = -i \times T/\sqrt{\frac{v^2}{c^2} - 1}$ or it can start clockwise because it is positive and time is said to be contracting since $T' = i \times T/\sqrt{\frac{v^2}{c^2} - 1}$ and this depends on the sign before the imaginary number i.

Accordingly, as we have noticed, the metauniverse is truly at a different level of experience, it is in fact beneath the atomic world when speaking about space (dimensions smaller than zero) and beyond infinity when speaking about time. In fact, we may ask where is this metauniverse if it is beyond infinity and beneath zero? The answer is evident and it is shown in the equations: in other dimensions, which form the meta-space-time of the metauniverse $\mathbf{G_2}$, in the world of the imaginary number i. If a new matter is indirectly detected (like dark matter) then Metarelativity is able to explain it and it takes into consideration its existence because no directly detectable and visible matter was found. Only its gravitational effect can be detected in $\mathbf{G_1}$. So, it should be another kind of matter, faster than light and unseen by our telescopes and accelerators. Thus, it should lay somewhere in space-time and this somewhere is the metauniverse $\mathbf{G_2}$. This will truly prove the existence of the metauniverse, which exists by mathematical and physical proofs and by the power of facts and experience. In fact, what is essentially more important in physics than the equations themselves is the understanding and the explanations given to these equations. What is more important than mathematics is its meaning and its philosophy.

4.6 The energy of the void

In fact, the metauniverse or the four imaginary dimensions of meta-space-time may be regarded as a field full of potential and latent energy as we have mentioned but "invisible" in nature since it is superluminal and imaginary as it is shown in the equation $E = \frac{\pm i m_0 \times c^2}{\sqrt{\frac{v^2}{c^2} - 1}} = E_{G2}$. We said a field because the hidden matter that lies inside it forms a field of action and potentialities that can be discovered, like in the atom. The metauniverse is a field of latent energy relative to $\mathbf{G_1}$. What does the metauniverse mean relatively to itself? The answer to this question was answered before in Metarelativity [1]. The result derives from the Metarelativistic equations. The outcome is that the metauniverse relatively to itself is just like the universe relatively to itself which means as real as the universe (refer to Section 4.7 in Chapter 1). I made the separation between both (between the two space-times) but in fact they are related and bonded both mathematically through precise equations. So, we have discovered the energy of the void, which is the invisible and dark and superluminal metauniverse $\mathbf{G_2}$ relative to the visible universe $\mathbf{G_1}$: its hidden imaginary dimensions, which lay in $\mathbf{G_2}$, its hidden energy (dark energy), and its hidden mass (dark matter).

4.7 Big Bang theory and the origin of the universe $\mathbf{G_1}$: Smaller than and before Planck's length and time (1.6×10^{-35} m and 10^{-43} seconds)

In fact, in our calculations, we expanded the Einstein-Lorentz equations to reach the metauniverse, as if we have done the backward walk by going from the universe $\mathbf{G_1}$ to the metauniverse $\mathbf{G_2}$. The direct walk is done from the metauniverse $\mathbf{G_2}$ to the universe $\mathbf{G_1}$ and is by saying that from this latent energy, that exists in the universe $\mathbf{G_2}$, the universe $\mathbf{G_1}$ emerged. In fact, if we do the direct walk, we will see our whole "real" universe $\mathbf{G_1}$ coming out from nothing, from void, from $\mathbf{G_2}$, to existence like in the Big Bang model. This "nothing" or this "void" that we noted is the superluminal imaginary metauniverse that we established its existence in Section 4.6. The dot or the geometric point (Section 4.5) that we were talking about is the singularity that general relativity talks about. In fact, according to the Big Bang model, from a singularity, all real space-time was generated and all matter within it. In the early fractions of a second, the particles and matter, the space-time itself were condensed in a small portion. This is said, we could assume that our visible universe came from another universe, which is the invisible metauniverse itself. This potent and latent energy or the energy of the void that represents the metauniverse comes from the invisible matter or the dark matter that is hidden in $\mathbf{G_2}$ in the total universe $\mathbf{G}$, which is denoted by:

$$\mathbf{G} = \text{universe } \mathbf{G_1} + \mathbf{EW} + \text{universe } \mathbf{G_2}.$$

which is similar to the complex set of numbers denoted in classical mathematics by $\mathbb{C}$. So, another proof of the existence of $\mathbf{G_2}$ is the Big Bang theory and that will be discussed more clearly and plainly in future publications.

4.8 Vacuum or the quantum field fluctuations and the uncertainty principle

Now, another proof of the existence of the metauniverse that will be demonstrated in forthcoming publications is at the level of the atom, where we have according to this uncertainty principle:

$$\Delta E \times \Delta t \geq \frac{h}{4\pi}$$

which is called the time-energy uncertainty relation also. The explanation of the principle of vacuum fluctuation is that energy is created from void during an interval Δt and then returns to void after creating virtual particles. In fact, the nothing or the vacuum as we have seen is the void that we have spoken about (Section 4.6) or the metauniverse $\mathbf{G_2}$ that we deduced from Metarelativity itself. Some physicists say that the whole universe is a quantum fluctuation phenomenon like in the principle of vacuum fluctuation. This is true if we looked at the equation from a different angle. If we reshape our minds, we may say that from the metauniverse a quantum phenomenon occurred that means a parcel of energy burst out from the metauniverse, that is full of potency, to "real existence," where the universe $\mathbf{G_1}$ was created and that will eventually disappear, say the physicists, in a period of time Δt, which is the age of the "real" universe $\mathbf{G_1}$.

Additionally, according to Metarelativity, the two complementary particles of metamatter $+i|m_{G2}|$ and $-i|m_{G2}|$ in the metauniverse $\mathbf{G_2}$ can annihilate into the real matter in the universe $\mathbf{G_1}$ or into photons in the universe $\mathbf{G_3}$. Therefore, vacuum fluctuation is nothing but the annihilations of metaparticles into the "real" universe $\mathbf{G_1}$ or into the luminal universe $\mathbf{G_3}$ (refer to Section 4.2 in Chapter 1) (**Figure 14**).

Accordingly, "Ex Nihilo Nihil Fit," or "Nothing Comes from Nothing" as argued by the Greek Parmenides.

4.9 Black holes as doors to the metauniverse G_2

Furthermore, and as a consequence of what has been said in the previous sections, we can understand that black holes that contain ultimately at their end the space-time singularities are nothing but doors to the metauniverse $\mathbf{G_2}$ (**Figure 15**).

4.10 Unification of the four interactions in G_1

Therefore, since the whole "real" universe $\mathbf{G_1}$ was created from the metauniverse $\mathbf{G_2}$ then all the four interactions in $\mathbf{G_1}$ were also created from $\mathbf{G_2}$. This is to say that

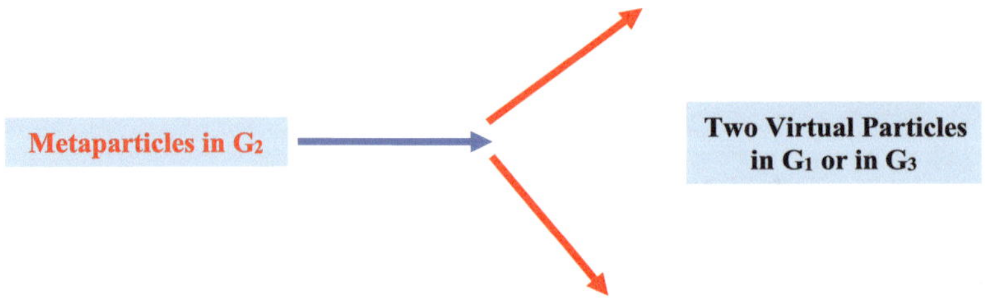

Figure 14.
Vacuum fluctuation and the uncertainty principle.

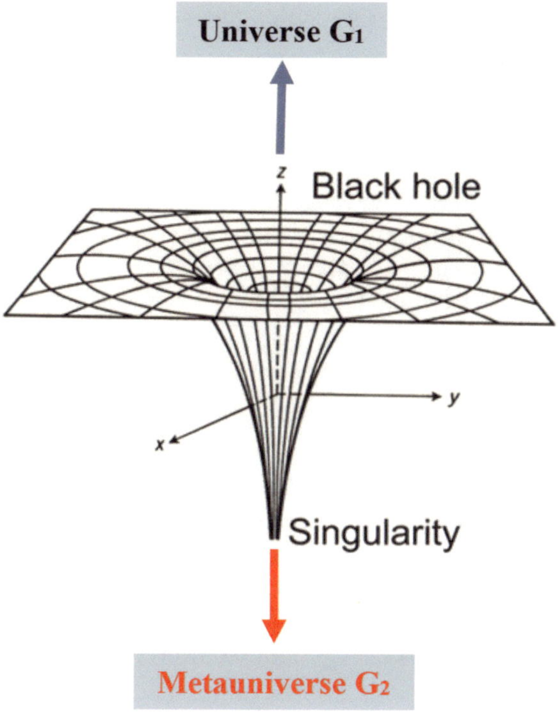

Figure 15.
Black holes as doors to the imaginary metauniverse $\mathbf{G_2}$.

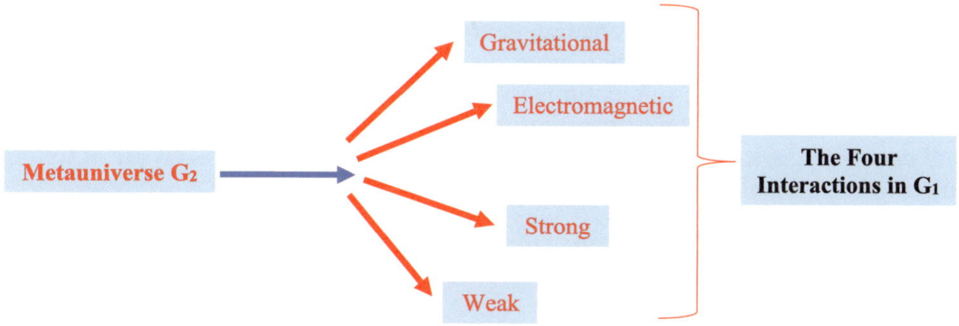

Figure 16.
Unification of the four interactions in the real universe $\mathbf{G_1}$.

nature four interactions that exist in $\mathbf{G_1}$ emerged from the metauniverse $\mathbf{G_2}$ just like the whole "real" universe $\mathbf{G_1}$ (**Figure 16**).

4.11 Ordinary matter/energy and positive/negative dark matter/energy

Sir Isaac Newton's law of gravitation of attraction is:

$F = k\frac{m \times m'}{d^2} = F_{G1}$ = the force in the universe $\mathbf{G_1}$, where k = $6.674 \times 10^{-11}\,\text{m}^3{\cdot}\text{kg}^{-1}{\cdot}\text{s}^{-2}$ approximately is the gravitational constant in the international system of units (SI).

$\Leftrightarrow F_G = k\frac{m_{G2} \times m_{G1}}{d^2} = k\frac{\pm im \times m'}{d^2} = \pm i\left(k\frac{m \times m'}{d^2}\right)$ in the universe $\mathbf{G}$.

So, as a consequence of the first possible solution is the Metarelativistic law of gravitation of attraction:

$$F_G = +i\left(k\frac{m \times m'}{d^2}\right)$$

⇒ Positive dark matter and energy of $\mathbf{G_2}$ can attract ordinary matter in $\mathbf{G_1}$.

⇒ Verification of astronomical observations and the explanation of the dark matter attraction of ordinary matter in galaxies, stars, etc.

And, as a consequence of the second possible solution is the Metarelativistic law of gravitation of repulsion:

$$F_G = -i\left(k\frac{m \times m'}{d^2}\right)$$

⇒ Negative dark matter and energy of $\mathbf{G_2}$ can repulse ordinary matter in $\mathbf{G_1}$.

⇒ Explanation of the expansion of the universe $\mathbf{G_1}$.

⇒ Verification of astronomical observations and the explanation of Einstein's general relativity cosmological constant Λ (his "biggest blunder").

4.12 Conservation of mass and energy and vacuum fluctuation

The Lavoisier principle in chemistry and science affirms that mass and energy are conserved. The Law of Conservation of Mass (or Matter) and Energy in a chemical reaction can be stated thus: In a chemical reaction, the matter is neither created nor destroyed.

"Nothing is lost, nothing is created, everything is transformed".

It was discovered by Antoine Laurent Lavoisier (1743–94) about 1785.

Knowing that vacuum fluctuations are the materialization or the annihilation or the transformation of metaparticles into real particles or photons, consequently, the total mass and the total energy in the total universe $\mathbf{G} = \mathbf{G_1} + \mathbf{EW} + \mathbf{G_2}$ are absolutely conserved, such that:

$$E_G = [E_{G1}] + [E_{G3}] + [E_{G2}] \Leftrightarrow E_G = [E_{G1}] + [\text{Energy of Electromagnetic waves}] + [\pm i|E_{G2}|]$$

And

$$m_G = [m_{G1}] + [m_{G3}] + [m_{G2}] \Leftrightarrow m_G = [m_{G1}] + [\text{Mass of Electromagnetic waves}] + [\pm i|m_{G2}|]$$

4.13 Determinism and nondeterminism

The mathematical probability concept was set forth by *Andrey* Nikolaevich *Kolmogorov* in 1933 by laying down a five-axioms system. This scheme has been improved in MCPP to embody the set of imaginary numbers after adding three new axioms. Accordingly, any stochastic phenomenon is performed in the probability set and total universe $\mathbf{G} = \mathcal{C}$ of complex probabilities, which is the summation of the set $\mathcal{R}$ of real probabilities and the set $\mathcal{M}$ of imaginary probabilities. Our objective in this work was to encompass complementary imaginary dimensions to the stochastic phenomenon taking place in the "real" probability laboratory in $\mathcal{R}$ and as a consequence to gauge in the sets $\mathcal{R}$, $\mathcal{M}$, and $\mathcal{C}$ all the corresponding probabilities. Hence, the probability in the entire set and total universe **G** = $\mathcal{C}$ = $\mathcal{R}$ + $\mathcal{M}$ is incessantly equal to one independently of all the probabilities of the input stochastic variable distribution in $\mathcal{R}$, and subsequently, the output of the random phenomenon in $\mathcal{R}$ can be evaluated totally and absolutely in **G** = $\mathcal{C}$. This is due to the fact that the probability in $\mathcal{C}$ is calculated after the elimination and subtraction of the chaotic factor from the degree of our knowledge of the nondeterministic phenomenon as it is shown in the equation: $Pc^2 = DOK - Chf = DOK + MChf = 1 = Pc$. Consequently, we have applied this novel CPP paradigm to an important and fundamental problem in physics, which is Metarelativity theory. Hence, and what is truly crucial, is that we have demonstrated that probabilistic phenomena can be expressed totally deterministically in the complex probability set and total universe **G** = $\mathcal{C} = \mathcal{R} + \mathcal{M}$. Therefore, and after all, "God does not play dice!!!" as Albert Einstein put it.

4.14 Entropy and metaentropy

To understand the meaning of negative time in $\mathbf{G_2}$ relatively to $\mathbf{G_1}$, then entropy is the best tool. We know that entropy is defined as $d[S_{G1}] \geq 0$ according to the second principle of thermodynamics. We say that when time grows, then entropy and chaos and disorder increase in $\mathbf{G_1}$. Due to the fact that time is negative as one possible solution in $\mathbf{G_2}$, this implies that we can have: $d[S_{G2}] \leq 0$. Consequently, and for this case, we say that when time flows, then entropy (or metaentropy) decreases. This means directly the following: The direction of evolution in a part of $\mathbf{G_2}$ is the opposite to that in $\mathbf{G_1}$.

Additionally, $d[S_G] = 0$ since $Ln(P_G) = Ln(Pc) = Ln(1) = 0$ in the expression of entropy in the theory of statistical mechanics:

$$S_G = -k_B \sum_j p_j Ln\left(p_j\right) = -k_B \sum_j 1 \times Ln(1) = 0.$$

That means and most importantly, that for any distribution and in the complex probability set and the total universe **G** = $\mathcal{C} = \mathcal{R} + \mathcal{M}$, we have complete order, no chaos, no ignorance, no uncertainty, no disorder, no randomness, no nondeterminism, and no unpredictability since all measurements are completely and perfectly and absolutely deterministic.

4.15 Conservation of information in G = $\mathcal{C}$

In the complex set and total universe **G** = $\mathcal{C}$ we have the entropy in the theory of information always equal to 0 since:

$$\Leftrightarrow H_G = -\sum_j p_j \, log_b\left(p_j\right) = -\sum_j 1 \times \, log_b(1) = 0$$

$$\Leftrightarrow d(H_G) = 0$$

So, no loss and no gain but complete conservation of information in **G**. Hence, the Lavoisier Law of Conservation of Mass and Energy applies also to information theory as it was shown here and in my previous published work [13, 15] so to as well *MCPP*. In $\mathcal{R}$, we have disorder, uncertainty, and unpredictability. In $\mathcal{C}$ we have order, certainty, and predictability since $Pc = 1$ permanently and entropy = 0 constantly. Additionally, in

$\mathcal{R}$ we have chaos and imperfect and incomplete knowledge or partial ignorance. In $\mathcal{C}$ we have chaos always equal to 0 ($Chf = 0$ and $MChf = 0$) and $DOK = 1$ continuously, thus complete and perfect and total knowledge of the now deterministic messages and the information of the now deterministic experiments occurring in $\mathbf{G} = \mathcal{C}$.

4.16 The duality and complementarity principle in $\mathcal{R}$, $\mathcal{M}$, and G

One of the fundamental principles almost omnipresent in whole nature and at all levels of experience in the physical world, including in some properties and in some characteristics of the atomic level, is duality and complementarity. This principle is present also in $\mathbf{G}_2$ where we have metaparticles with their corresponding metaEnergies and metaAntiparticles with their corresponding metaAntiEnergies. Also, the probability P_m in $\mathcal{M}$ is the associated imaginary complement of the real probability P_r in $\mathcal{R}$, their sum in $\mathcal{C}$ is $Pc = 1$. Additionally, and in **G**, the superluminal metauniverse $\mathbf{G}_2$ is the "imaginary" complement of the "real" subluminal universe $\mathbf{G}_1$. Hence, we can directly see that this principle dominates nearly all existence whether in $\mathcal{R}$, $\mathcal{M}$, or $\mathbf{G} = \mathcal{C}$.

5. Conclusion and perspectives

In the current research work, the original extended model of eight axioms (*EKA*) of A. N. Kolmogorov was connected and applied to Metarelativity theory. Thus, a tight link between Metarelativity and the novel paradigm (*CPP*) was achieved. Consequently, the model of "Complex Probability" was more developed beyond the scope of my 21 previous research works on this topic.

Furthermore, the theory of Metarelativity is a system of equations written to take into consideration additional effects in the universe and about the matter inside it. Metarelativity begins with Albert Einstein's theory of special relativity and it develops a system of equations that lead us to further explanations and to a new physics paradigm. Like special relativity which was created in 1905 and then expanded later to general relativity to explain, among other things, the aberration in the motion of the planet Mercury and the gravitational lenses, Metarelativity explains many phenomena, for example, the nature of dark matter laying inside and outside galaxies and, in the universe, and the existence of superluminal particles or tachyons and their corresponding dark energy. Metarelativity is a work of pure science that encompasses mathematics and fundamental physics. All the explanations are deduced from a new system of equations called the Metarelativistic transformations that were proven mathematically and explained physically. Hence, Metarelativity was bonded here to *CPP* and to develop a new paradigm in science.

In addition, referring to all these obtained graphs and executed simulations throughout the whole research work, we are able to quantify and visualize both the system chaos and stochastic effects and influences (expressed and materialized by *Chf* and *MChf*) and the certain knowledge (expressed and materialized by *DOK* and *Pc*) of the new paradigm. This is without any doubt very fruitful, wonderful, and fascinating and proves and reveals once again the advantages of extending A. N. Kolmogorov's five axioms of probability and hence the novelty and benefits of my inventive and original model in the fields of prognostics, applied mathematics, and physics that can be called verily: "The Metarelativistic Complex Probability Paradigm (*MCPP*)."

As a future and prospective research and challenges, we aim to more develop the novel prognostic paradigm conceived and to implement it to a large set of random and nondeterministic phenomena in physics and in science.

Author details

Abdo Abou Jaoudé
Faculty of Natural and Applied Sciences, Department of Mathematics and Statistics, Notre Dame University-Louaize, Lebanon

*Address all correspondence to: abdoaj@idm.net.lb

References

[1] Abou Jaoude A. The theory of Metarelativity: Beyond Albert Einstein's relativity. Physics International, Science Publications. 2013;**4**(2):97-109

[2] Wikipedia, the Free Encyclopedia, Special Relativity. Available from: https://en.wikipedia.org/

[3] Wikipedia, the Free Encyclopedia, General Relativity. Available from: https://en.wikipedia.org/

[4] Abou Jaoude A, El-Tawil K, Kadry S. Prediction in complex dimension using Kolmogorov's set of axioms. Journal of Mathematics and Statistics, Science Publications. 2010;**6**(2):116-124

[5] Abou Jaoude A. The complex statistics paradigm and the law of large numbers. Journal of Mathematics and Statistics, Science Publications. 2013;**9**(4): 289-304

[6] Abou Jaoude A. The theory of complex probability and the first order reliability method. Journal of Mathematics and Statistics, Science Publications. 2013;**9**(4):310-324

[7] Abou Jaoude A. Complex probability theory and prognostic. Journal of Mathematics and Statistics, Science Publications. 2014;**10**(1):1-24

[8] Abou Jaoude A. The complex probability paradigm and analytic linear prognostic for vehicle suspension systems. American Journal of Engineering and Applied Sciences, Science Publications. 2015;**8**(1):147-175

[9] Abou Jaoude A. The paradigm of complex probability and the Brownian motion. Systems Science and Control Engineering, Taylor and Francis Publishers. 2015;**3**(1):478-503

[10] Abou Jaoude A. The paradigm of complex probability and Chebyshev's inequality. Systems Science and Control Engineering, Taylor and Francis Publishers. 2016;**4**(1):99-137

[11] Abou Jaoude A. The paradigm of complex probability and analytic nonlinear prognostic for vehicle suspension systems. Systems Science and Control Engineering, Taylor and Francis Publishers. 2016;**4**(1):99-137

[12] Abou Jaoude A. The paradigm of complex probability and analytic linear prognostic for unburied petrochemical pipelines. Systems Science and Control Engineering, Taylor and Francis Publishers. 2017;**5**(1): 178-214

[13] Abou Jaoude A. The paradigm of complex probability and Claude Shannon's information theory. Systems Science and Control Engineering, Taylor and Francis Publishers. 2017;**5**(1): 380-425

[14] Abou Jaoude A. The paradigm of complex probability and analytic nonlinear prognostic for unburied petrochemical pipelines. Systems Science and Control Engineering, Taylor and Francis Publishers. 2017;**5**(1):495-534

[15] Abou Jaoude A. The paradigm of complex probability and Ludwig Boltzmann's entropy. Systems Science and Control Engineering, Taylor and Francis Publishers. 2018;**6**(1): 108-149

[16] Abou Jaoude A. The paradigm of complex probability and Monte Carlo methods. Systems Science and Control Engineering, Taylor and Francis Publishers. 2019;**7**(1):407-451

[17] Abou Jaoude A. Analytic prognostic in the linear damage case applied to buried petrochemical pipelines and the complex probability paradigm. Fault Detection, Diagnosis and Prognosis, IntechOpen. 2020;**1**(5):65-103. DOI: 10.5772/intechopen.90157

[18] Abou Jaoude A. The Monte Carlo techniques and the complex probability paradigm. Forecasting in Mathematics - Recent Advances, New Perspectives and Applications, IntechOpen. 2020;**1**(1):1-29. DOI: 10.5772/intechopen.93048

[19] Abou Jaoude A. The paradigm of complex probability and prognostic using FORM. London Journal of Research in Science: Natural and Formal (LJRS). 2020;**20**(4):1-65. London Journals Press, Print ISSN: 2631-8490, Online ISSN: 2631-8504. DOI: 10.17472/LJRS

[20] Abou Jaoude A. The paradigm of complex probability and the central limit theorem. London Journal of Research in Science: Natural and Formal (LJRS). 2020;**20**(5):1-57. London Journals Press, Print ISSN: 2631-8490, Online ISSN: 2631-8504. DOI: 10.17472/LJRS

[21] Abou Jaoude A. The paradigm of complex probability and Thomas Bayes' theorem. The Monte Carlo Methods - Recent Advances, New Perspectives and Applications. IntechOpen; **1**(1):1-44. ISBN: 978-1-83968-760-0. 2021. DOI: 10.5772/intechopen.98340

[22] Abou Jaoude A. The paradigm of complex probability and Isaac Newton's classical mechanics: On the Foundation of Statistical Physics. The Monte Carlo Methods - Recent Advances, New Perspectives and Applications. 2021;**1**(2):45-116. IntechOpen. ISBN: 978-1-83968-760-0. DOI: 10.5772/intechopen.98341

[23] Abou Jaoude A. The paradigm of complex probability and quantum mechanics: The infinite potential well problem - the position wave function. Applied Probability Theory - New Perspectives, Recent Advances and Trends. 2022;**1**(1):1-44. IntechOpen. DOI: 10.5772/intechopen.107300

[24] Abou Jaoude A. The paradigm of complex probability and quantum mechanics: The infinite potential well problem – The momentum Wavefunction and the Wavefunction entropies. Applied Probability Theory – New Perspectives, Recent Advances and Trends. 2022. IntechOpen. 2022;**1**(2):45-88. DOI: 10.5772/intechopen.107665

[25] Benton W. Probability, Encyclopedia Britannica. Vol. 18. Chicago: Encyclopedia Britannica Inc; 1966. pp. 570-574

[26] Benton W. Mathematical Probability, Encyclopedia Britannica. Vol. 18. Chicago: Encyclopedia Britannica Inc; 1966. pp. 574-579

[27] Feller W. An Introduction to Probability Theory and its Applications. 3rd ed. New York: Wiley; 1968

[28] Walpole R, Myers R, Myers S, Ye K. Probability and Statistics for Engineers and Scientists. 7th ed. New Jersey: Prentice Hall; 2002

[29] Freund JE. Introduction to Probability. New York: Dover Publications; 1973

[30] Abou Jaoude A. The Computer Simulation of Monté Carlo Methods and Random Phenomena. United Kingdom: Cambridge Scholars Publishing; 2019

[31] Abou Jaoude A. The Analysis of Selected Algorithms for the Stochastic

Paradigm. United Kingdom: Cambridge Scholars Publishing; 2019

[32] Abou Jaoude A. The Analysis of Selected Algorithms for the Statistical Paradigm. Vol. 1. The Republic of Moldova: Generis Publishing; 2021

[33] Abou Jaoude A. The Analysis of Selected Algorithms for the Statistical Paradigm. Vol. 2. The Republic of Moldova: Generis Publishing; 2021

[34] Abou Jaoude A. Forecasting in Mathematics - Recent Advances, New Perspectives and Applications. London, UK, London, United Kingdom: IntechOpen; 2021. ISBN: 978-1-83880-825-9

[35] Abou Jaoude A. The Monte Carlo Methods - Recent Advances, New Perspectives and Applications. London, UK, London, United Kingdom: IntechOpen; 2022. ISBN: 978-1-83968-760-0

[36] Abou Jaoude A. Applied Probability Theory – New Perspectives, Recent Advances and Trends. London, UK, United Kingdom: IntechOpen. London; 2023. ISBN: 978-1-83768-296-6

[37] Abou Jaoude A. Ph.D. Thesis in Applied Mathematics: Numerical Methods and Algorithms for Applied Mathematicians. Spain: Bircham International University; 2004. Available from: http://www.bircham.edu

[38] Abou Jaoude A. Ph.D. Thesis in Computer Science: Computer Simulation of Monté Carlo Methods and Random Phenomena. Spain: Bircham International University; 2005. Available from: http://www.bircham.edu

[39] Abou Jaoude A. Ph.D. Thesis in Applied Statistics and Probability: Analysis and Algorithms for the Statistical and Stochastic Paradigm. Spain: Bircham International University; 2007. Available from: http://www.bircham.edu

[40] Wikipedia, the Free Encyclopedia, Dark Matter. Available from: https://en.wikipedia.org/

[41] Wikipedia, the Free Encyclopedia, Dark Energy. Available from: https://en.wikipedia.org/

[42] Wikipedia, the Free Encyclopedia, Lambda-CDM Model. https://en.wikipedia.org

[43] Heath TL. The Elements of Euclid. New York, USA: Dover; 1956

Chapter 3

Sets of Fractional Operators and Some of Their Applications

A. Torres-Hernandez, F. Brambila-Paz and R. Ramirez-Melendez

Abstract

This chapter presents one way to define Abelian groups of fractional operators isomorphic to the group of integers under addition through a family of sets of fractional operators and a modified Hadamard product, as well as one way to define finite Abelian groups of fractional operators through sets of positive residual classes less than a prime number. Furthermore, it is presented one way to define sets of fractional operators which allow generalizing the Taylor series expansion of a vector-valued function in multi-index notation, as well as one way to define a family of fractional fixed-point methods and determine their order of convergence analytically through sets.

Keywords: fractional operators, set theory, group theory, fractional iterative methods, fractional calculus of sets

1. Introduction

In one dimension, a fractional derivative may be considered in a general way as a parametric operator of order α, such that it coincides with conventional derivatives when α is a positive integer n. So, when it is not necessary to explicitly specify the form of a fractional derivative, it is usually denoted as follows

$$\frac{d^\alpha}{dx^\alpha}. \tag{1}$$

On the other hand, a fractional differential equation is an equation that involves at least one differential operator of order α, with $(n-1) < \alpha \leq n$ for some positive integer n, and it is said to be a differential equation of order α if this operator is the highest order in the equation. The fractional operators have many representations [1–3], but one of their fundamental properties is that they allow retrieving the results of conventional calculus when $\alpha \to n$. For example, let $f : \Omega \subset \mathbb{R} \to \mathbb{R}$ be a function such that $f \in L^1_{loc}(a, b)$, where $L^1_{loc}(a, b)$ denotes the space of locally integrable functions on the open interval $(a, b) \subset \Omega$. One of the fundamental operators of fractional calculus is the operator Riemann–Liouville fractional integral, which is defined as follows [4, 5]:

$${}_aI_x^\alpha f(x) := \frac{1}{\Gamma(\alpha)} \int_a^x (x-t)^{\alpha-1} f(t)dt, \tag{2}$$

where Γ denotes the Gamma function. It is worth mentioning that the above operator is a fundamental piece to construct the operator Riemann–Liouville fractional derivative, which is defined as follows [4, 6]:

$$ {}_aD_x^\alpha f(x) := \begin{cases} {}_aI_x^{-\alpha} f(x), & \text{if } \alpha < 0 \\ \dfrac{d^n}{dx^n}\left({}_aI_x^{n-\alpha} f(x)\right), & \text{if } \alpha \geq 0 \end{cases}, \tag{3} $$

where $n = \lceil \alpha \rceil$ and ${}_aI_x^0 f(x) := f(x)$. On the other hand, let $f : \Omega \subset \mathbb{R} \to \mathbb{R}$ be a function n-times differentiable such that $f, f^{(n)} \in L^1_{loc}(a, b)$. Then, the Riemann–Liouville fractional integral also allows constructing the operator Caputo fractional derivative, which is defined as follows [4, 6]:

$$ {}_a^C D_x^\alpha f(x) := \begin{cases} {}_aI_x^{-\alpha} f(x), & \text{if } \alpha < 0 \\ {}_aI_x^{n-\alpha} f^{(n)}(x), & \text{if } \alpha \geq 0 \end{cases}, \tag{4} $$

where $n = \lceil \alpha \rceil$ and ${}_aI_x^0 f^{(n)}(x) := f^{(n)}(x)$. Furthermore, if the function f fulfills that $f^{(k)}(a) = 0 \ \forall k \in \{0, 1, \cdots, n-1\}$, the Riemann–Liouville fractional derivative coincides with the Caputo fractional derivative, that is,

$$ {}_aD_x^\alpha f(x) = {}_a^C D_x^\alpha f(x). \tag{5} $$

So, applying the operator (3) with $a = 0$ to the function x^μ, with $\mu > -1$, we obtain the following result [7]:

$$ {}_0D_x^\alpha x^\mu = \frac{\Gamma(\mu+1)}{\Gamma(\mu-\alpha+1)} x^{\mu-\alpha}, \quad \alpha \in \mathbb{R} \backslash \mathbb{Z}, \tag{6} $$

where if $1 \leq \lceil \alpha \rceil \leq \mu$ it is fulfilled that ${}_0D_x^\alpha x^\mu = {}_0^C D_x^\alpha x^\mu$.

2. Sets of fractional operators

Before continuing, it is necessary to mention that due to the large number of fractional operators that may exist [1–3, 8–23], some sets must be defined to fully characterize elements of fractional calculus. It is worth mentioning that characterizing elements of fractional calculus through sets is the main idea behind of the methodology known as fractional calculus of sets [24, 25]. So, considering a scalar function $h : \mathbb{R}^m \to \mathbb{R}$ and the canonical basis of $\mathbb{R}^m$ denoted by $\{\hat{e}_k\}_{k \geq 1}$, it is possible to define the following fractional operator of order α using Einstein notation

$$ o_x^\alpha h(x) := \hat{e}_k o_k^\alpha h(x). \tag{7} $$

Therefore, denoting by ∂_k^n the partial derivative of order n applied with respect to the k-th component of the vector x, using the previous operator it is possible to define the following set of fractional operators

$$ O_{x,\alpha}^n(h) := \left\{ o_x^\alpha \ : \ \exists o_k^\alpha h(x) \quad \text{and} \quad \lim_{\alpha \to n} o_k^\alpha h(x) = \partial_k^n h(x) \ \forall k \geq 1 \right\}, \tag{8} $$

which may be proved to be a nonempty set through the following set of fractional operators

$$O^n_{0,x,\alpha}(h) := \left\{ o^\alpha_x \ : \ \exists o^\alpha_k h(x) = \left(\partial^n_k + \mu(\alpha)\partial^\alpha_k\right)h(x) \quad \text{and} \quad \lim_{\alpha \to n} \mu(\alpha)\partial^\alpha_k h(x) = 0 \ \ \forall k \geq 1 \right\}, \tag{9}$$

with which it is possible to obtain the following result:

$$\text{If} \quad o^\alpha_{i,x}, o^\alpha_{j,x} \in O^n_{x,\alpha}(h) \quad \text{with} \quad i \neq j \quad \Rightarrow \quad \exists o^\alpha_{k,x} = \frac{1}{2}\left(o^\alpha_{i,x} + o^\alpha_{j,x}\right) \in O^n_{x,\alpha}(h). \tag{10}$$

So, the complement of the set (8) may be defined as follows

$$O^{n,c}_{x,\alpha}(h) := \left\{ o^\alpha_x \ : \exists o^\alpha_k h(x) \quad \forall k \geq 1 \quad \text{and} \quad \lim_{\alpha \to n} o^\alpha_k h(x) \neq \partial^n_k h(x) \quad \text{in} \right. \tag{11}$$

$$\left. \text{at least one value} \quad k \geq 1 \right\},$$

with which it is possible to obtain the following result:

$$\text{If} \quad o^\alpha_{i,x} = \hat{e}_k o^\alpha_{i,k} \in O^n_{x,\alpha}(h) \quad \Rightarrow \quad \exists o^\alpha_{j,x} = \hat{e}_k o^\alpha_{i,\sigma_j(k)} \in O^{n,c}_{x,\alpha}(h), \tag{12}$$

where $\sigma_j : \{1, 2, \cdots, m\} \to \{1, 2, \cdots, m\}$ denotes any permutation different from the identity. On the other hand, the set (8) may be considered as a generating set of sets of fractional tensor operators. For example, considering $\alpha, n \in \mathbb{R}^d$ with $\alpha = \hat{e}_k[\alpha]_k$ and $n = \hat{e}_k[n]_k$, it is possible to define the following set of fractional tensor operators

$$O^n_{x,\alpha}(h) := \left\{ o^\alpha_x \ : \ \exists o^\alpha_x h(x) \quad \text{and} \quad o^\alpha_x \in O^{[n]_1}_{x,[\alpha]_1}(h) \times O^{[n]_2}_{x,[\alpha]_2}(h) \times \cdots \times O^{[n]_d}_{x,[\alpha]_d}(h) \right\}. \tag{13}$$

3. Groups of fractional operators

Considering a function $h : \Omega \subset \mathbb{R}^m \to \mathbb{R}^m$, it is possible to define the following sets

$${}_m\mathrm{O}^n_{x,\alpha}(h) := \left\{ o^\alpha_x \ : \ o^\alpha_x \in \mathrm{O}^n_{x,\alpha}\left([h]_k\right) \ \ \forall k \leq m \right\}, \tag{14}$$

$${}_m\mathrm{O}^{n,c}_{x,\alpha}(h) := \left\{ o^\alpha_x \ : \ o^\alpha_x \in \mathrm{O}^{n,c}_{x,\alpha}\left([h]_k\right) \ \ \forall k \leq m \right\}, \tag{15}$$

$${}_m\mathrm{O}^{n,u}_{x,\alpha}(h) := {}_m\mathrm{O}^n_{x,\alpha}(h) \cup {}_m\mathrm{O}^{n,c}_{x,\alpha}(h), \tag{16}$$

where $[h]_k : \Omega \subset \mathbb{R}^m \to \mathbb{R}$ denotes the k-th component of the function h. So, it is possible to define the following set of fractional operators

$${}_m\mathrm{MO}^{\infty,u}_{x,\alpha}(h) := \bigcap_{k \in \mathbb{Z}} m\mathrm{O}^{k,u}_{x,\alpha}(h), \tag{17}$$

which under the classical Hadamard product it is fulfilled that

$$o^0_x \circ h(x) := h(x) \quad \forall o^\alpha_x \in {}_m\mathrm{MO}^{\infty,u}_{x,\alpha}(h). \tag{18}$$

As a consequence, it is possible to define the following set of matrices

$$ {}_m\mathrm{M}^{\infty}_{x,\alpha}(h) := \left\{ A_{h,\alpha} = A_{h,\alpha}\left(o^{\alpha}_{x}\right) : o^{\alpha}_{x} \in {}_m\mathrm{MO}^{\infty,u}_{x,\alpha}(h) \text{ and } A_{h,\alpha}(x) = \left([A_{h,\alpha}]_{jk}(x) \right) := \left(o^{\alpha}_{k}[h]_j(x) \right) \right\}, \tag{19} $$

and therefore, considering that when using the classical Hadamard product in general $o^{p\alpha}_{x} \circ o^{q\alpha}_{x} \neq o^{(p+q)\alpha}_{x}$. Assuming the existence of a fixed set of matrices ${}_m\mathrm{M}^{\infty}_{x,\alpha}(h)$, joined with a modified Hadamard product that fulfills the following property

$$ o^{p\alpha}_{i,x} \circ o^{q\alpha}_{j,x} := \begin{cases} o^{p\alpha}_{i,x} \circ o^{q\alpha}_{j,x}, & \text{if } i \neq j \quad \text{(Hadamard product of type horizontal)} \\ o^{(p+q)\alpha}_{i,x}, & \text{if } i = j \quad \text{(Hadamard product of type vertical)} \end{cases}, \tag{20} $$

by omitting the function h, the resulting set ${}_m\mathrm{M}^{\infty}_{x,\alpha}(\cdot)$ has the ability to generate a group of fractional matrix operators A_α that fulfill the following equation

$$ A_\alpha\left(o^{p\alpha}_{i,x}\right) \circ A_\alpha\left(o^{q\alpha}_{j,x}\right) := \begin{cases} A_\alpha\left(o^{p\alpha}_{i,x} \circ o^{q\alpha}_{j,x}\right), & \text{if } i \neq j \\ A_\alpha\left(o^{(p+q)\alpha}_{i,x}\right), & \text{if } i = j \end{cases}, \tag{21} $$

through the following set [24, 26]:

$$ {}_m\mathrm{G}_{FIM}(\alpha) := \left\{ A^{\circ r}_{\alpha} = A_\alpha\left(o^{r\alpha}_{x}\right) : \exists A^{\circ r}_{\alpha} \in {}_m\mathrm{M}^{\infty}_{x,\alpha}(\cdot) \ \forall r \in \mathbb{Z} \text{ and } A^{\circ r}_{\alpha} = \left(\left[A^{\circ r}_{\alpha}\right]_{jk} \right) := \left(o^{r\alpha}_{k} \right) \right\}. \tag{22} $$

Where $\forall A^{\circ p}_{i,\alpha}, A^{\circ q}_{j,\alpha}, A^{\circ r}_{j,\alpha} \in {}_m\mathrm{G}_{FNR}(\alpha)$, with $i \neq j$, the following property is defined

$$ \left(A^{\circ p}_{i,\alpha} \circ A^{\circ q}_{j,\alpha}\right) \circ A^{\circ r}_{j,\alpha} = A^{\circ p}_{i,\alpha} \circ \left(A^{\circ q}_{j,\alpha} \circ A^{\circ r}_{j,\alpha}\right) = A^{\circ 1}_{k,\alpha} := A_{k,\alpha}\left(o^{p\alpha}_{i,x} \circ o^{(q+r)\alpha}_{j,x}\right), \quad p,q,r \in \mathbb{Z} \setminus \{0\}, \tag{23} $$

since it is considered that through combinations of the Hadamard product of type horizontal and vertical the fractional operators are reduced to their minimal expression. As a consequence, it is fulfilled that

$$ \begin{aligned} &\forall A^{\circ 1}_{k,\alpha} \in {}_m\mathrm{G}_{FIM}(\alpha) \quad \text{such that} \quad A_{k,\alpha}\left(o^{\alpha}_{k,x}\right) = A_{k,\alpha}\left(o^{p\alpha}_{i,x} \circ o^{q\alpha}_{j,x}\right) \exists A^{\circ r}_{k,\alpha} = A^{\circ (r-1)}_{k,\alpha} \circ A^{\circ 1}_{k,\alpha} \\ &\quad = A_{k,\alpha}\left(o^{rp\alpha}_{i,x} \circ o^{rq\alpha}_{j,x}\right). \end{aligned} \tag{24} $$

It is necessary to mention that for each operator $o^{\alpha}_{x} \in {}_m\mathrm{MO}^{\infty,u}_{x,\alpha}(h)$ it is possible to define a group [26], which is isomorphic to the group of integers under the addition, as shown by the following theorems:

Theorem 1.1 Let o^{α}_{x} be a fractional operator such that $o^{\alpha}_{x} \in {}_m\mathrm{MO}^{\infty,u}_{x,\alpha}(h)$. So, considering the modified Hadamard product given by (20), it is possible to define the following set of fractional matrix operators

$$ {}_{m}G\left(A_{\alpha}\left(o_{x}^{\alpha}\right)\right) := \left\{A_{\alpha}^{\circ r} = A_{\alpha}\left(o_{x}^{r\alpha}\right) \quad : \quad r \in \mathbb{Z} \quad \text{and} \quad A_{\alpha}^{\circ r} = \left(\left[A_{\alpha}^{\circ r}\right]_{jk}\right) := \left(o_{k}^{r\alpha}\right)\right\}, \tag{25} $$

which corresponds to the Abelian group generated by the operator $A_{\alpha}\left(o_{x}^{\alpha}\right)$.

Proof: It should be noted that due to the way the set (25) is defined, just the Hadamard product of type vertical is applied among its elements. So, $\forall A_{\alpha}^{\circ p}, A_{\alpha}^{\circ q} \in {}_{m}G\left(A_{\alpha}\left(o_{x}^{\alpha}\right)\right)$ it is fulfilled that

$$ A_{\alpha}^{\circ p} \circ A_{\alpha}^{\circ q} = \left(\left[A_{\alpha}^{\circ p}\right]_{jk}\right) \circ \left(\left[A_{\alpha}^{\circ q}\right]_{jk}\right) = \left(o_{k}^{(p+q)\alpha}\right) = \left(\left[A_{\alpha}^{\circ (p+q)}\right]_{jk}\right) = A_{\alpha}^{\circ (p+q)}, \tag{26} $$

with which it is possible to prove that the set ${}_{m}G\left(A_{\alpha}\left(o_{x}^{\alpha}\right)\right)$ fulfills the following properties, which correspond to the properties of an Abelian group:

$$ \begin{cases} \forall A_{\alpha}^{\circ p}, A_{\alpha}^{\circ q}, A_{\alpha}^{\circ r} \in {}_{m}G\left(A_{\alpha}\left(o_{x}^{\alpha}\right)\right) \text{ it is fulfilled that } \left(A_{\alpha}^{\circ p} \circ A_{\alpha}^{\circ q}\right) \circ A_{\alpha}^{\circ r} = A_{\alpha}^{\circ p} \circ \left(A_{\alpha}^{\circ q} \circ A_{\alpha}^{\circ r}\right) 0.1cm \\ \exists A_{\alpha}^{\circ 0} \in {}_{m}G\left(A_{\alpha}\left(o_{x}^{\alpha}\right)\right) \text{ such that } \forall A_{\alpha}^{\circ p} \in {}_{m}G\left(A_{\alpha}\left(o_{x}^{\alpha}\right)\right) \text{ it is fulfilled that } A_{\alpha}^{\circ 0} \circ A_{\alpha}^{\circ p} = A_{\alpha}^{\circ p} 0.1cm \\ \forall A_{\alpha}^{\circ p} \in {}_{m}G\left(A_{\alpha}\left(o_{x}^{\alpha}\right)\right) \ \exists A_{\alpha}^{\circ -p} \in {}_{m}G\left(A_{\alpha}\left(o_{x}^{\alpha}\right)\right) \text{ such that } A_{\alpha}^{\circ p} \circ A_{\alpha}^{\circ -p} = A_{\alpha}^{\circ 0} 0.1cm \\ \forall A_{\alpha}^{\circ p}, A_{\alpha}^{\circ q} \in {}_{m}G\left(A_{\alpha}\left(o_{x}^{\alpha}\right)\right) \text{ it is fulfilled that } A_{\alpha}^{\circ p} \circ A_{\alpha}^{\circ q} = A_{\alpha}^{\circ q} \circ A_{\alpha}^{\circ p} \end{cases} . \tag{27} $$

Theorem 1.2 Let o_{x}^{α} be a fractional operator such that $o_{x}^{\alpha} \in {}_{m}\mathrm{MO}_{x,\alpha}^{\infty,u}(h)$ and let $(\mathbb{Z}, +)$ be the group of integers under the addition. So, the group generated by the operator $A_{\alpha}\left(o_{x}^{\alpha}\right)$ is isomorphic to the group $(\mathbb{Z}, +)$, that is,

$$ {}_{m}G\left(A_{\alpha}\left(o_{x}^{\alpha}\right)\right) \cong (\mathbb{Z}, +). \tag{28} $$

Proof: To prove the theorem it is enough to define a bijective homomorphism between the sets ${}_{m}G\left(A_{\alpha}\left(o_{x}^{\alpha}\right)\right)$ and $(\mathbb{Z}, +)$. Let $\psi : {}_{m}G\left(A_{\alpha}\left(o_{x}^{\alpha}\right)\right) \to (\mathbb{Z}, +)$ be a function with inverse function $\psi^{-1} : (\mathbb{Z}, +) \to {}_{m}G\left(A_{\alpha}\left(o_{x}^{\alpha}\right)\right)$. So, the functions ψ and ψ^{-1} may be defined as follows

$$ \psi\left(A_{\alpha}^{\circ r}\right) = r \quad \text{and} \quad \psi^{-1}(r) = A_{\alpha}^{\circ r}, \tag{29} $$

with which it is possible to obtain the following results:

$$ \begin{cases} \forall A_{\alpha}^{\circ p}, A_{\alpha}^{\circ q} \in {}_{m}G\left(A_{\alpha}\left(o_{x}^{\alpha}\right)\right) \text{ it is fulfilled that } \psi\left(A_{\alpha}^{\circ p} \circ A_{\alpha}^{\circ q}\right) = \psi\left(A_{\alpha}^{\circ (p+q)}\right) = p + q = \psi\left(A_{\alpha}^{\circ p}\right) + \psi\left(A_{\alpha}^{\circ q}\right) \\ \forall p, q \in (\mathbb{Z}, +) \text{ it is fulfilled that } \psi^{-1}(p+q) = A_{\alpha}^{\circ (p+q)} = A_{\alpha}^{\circ p} \circ A_{\alpha}^{\circ q} = \psi^{-1}(p) \circ \psi^{-1}(q) \end{cases} . \tag{30} $$

Therefore, from the previous results, it follows that the function ψ defines an isomorphism between the sets ${}_{m}G\left(A_{\alpha}\left(o_{x}^{\alpha}\right)\right)$ and $(\mathbb{Z}, +)$.

Then, from the previous theorems it is possible to obtain the following corollaries:

Corollary 1.3 *Let o_{x}^{α} be a fractional operator such that $o_{x}^{\alpha} \in {}_{m}\mathrm{MO}_{x,\alpha}^{\infty,u}(h)$ and let $(\mathbb{Z}, +)$ be the group of integers under the addition. So, considering the modified Hadamard product given by (20) and some subgroup $\mathbb{H}$ of the group $(\mathbb{Z}, +)$, it is possible to define the following set of fractional matrix operators*

$$ {}_mG\big(A_\alpha(o_x^\alpha),\, \mathbb{H}\big) := \left\{ A_\alpha^{\circ r} = A_\alpha\big(o_x^{r\alpha}\big) \;:\; r \in \mathbb{H} \quad \text{and} \quad A_\alpha^{\circ r} = \left(\left[A_\alpha^{\circ r}\right]_{jk} \right) := \left(o_k^{r\alpha}\right) \right\}, \tag{31} $$

which corresponds to a subgroup of the group generated by the operator $A_\alpha(o_x^\alpha)$, that is,

$$ {}_mG\big(A_\alpha(o_x^\alpha),\, \mathbb{H}\big) \leq {}_mG\big(A_\alpha(o_x^\alpha)\big). \tag{32} $$

Example 1 *Let $\mathbb{Z}_n$ be the set of residual classes less than n. So, considering a fractional operator $o_x^\alpha \in {}_m\mathrm{MO}_{x,\alpha}^{\infty,u}(h)$ and the set $\mathbb{Z}_{14}$, it is possible to define the Abelian group of fractional matrix operators ${}_mG\big(A_\alpha(o_x^\alpha),\, \mathbb{Z}_{14}\big)$. Furthermore, all possible combinations of the elements of the group under the modified Hadamard product given by (20) are summarized below:*

$\circ$	$A_\alpha^{\circ 0}$	$A_\alpha^{\circ 1}$	$A_\alpha^{\circ 2}$	$A_\alpha^{\circ 3}$	$A_\alpha^{\circ 4}$	$A_\alpha^{\circ 5}$	$A_\alpha^{\circ 6}$	$A_\alpha^{\circ 7}$	$A_\alpha^{\circ 8}$	$A_\alpha^{\circ 9}$	$A_\alpha^{\circ 10}$	$A_\alpha^{\circ 11}$	$A_\alpha^{\circ 12}$	$A_\alpha^{\circ 13}$
$A_\alpha^{\circ 0}$	$A_\alpha^{\circ 0}$	$A_\alpha^{\circ 1}$	$A_\alpha^{\circ 2}$	$A_\alpha^{\circ 3}$	$A_\alpha^{\circ 4}$	$A_\alpha^{\circ 5}$	$A_\alpha^{\circ 6}$	$A_\alpha^{\circ 7}$	$A_\alpha^{\circ 8}$	$A_\alpha^{\circ 9}$	$A_\alpha^{\circ 10}$	$A_\alpha^{\circ 11}$	$A_\alpha^{\circ 12}$	$A_\alpha^{\circ 13}$
$A_\alpha^{\circ 1}$	$A_\alpha^{\circ 1}$	$A_\alpha^{\circ 2}$	$A_\alpha^{\circ 3}$	$A_\alpha^{\circ 4}$	$A_\alpha^{\circ 5}$	$A_\alpha^{\circ 6}$	$A_\alpha^{\circ 7}$	$A_\alpha^{\circ 8}$	$A_\alpha^{\circ 9}$	$A_\alpha^{\circ 10}$	$A_\alpha^{\circ 11}$	$A_\alpha^{\circ 12}$	$A_\alpha^{\circ 13}$	$A_\alpha^{\circ 0}$
$A_\alpha^{\circ 2}$	$A_\alpha^{\circ 2}$	$A_\alpha^{\circ 3}$	$A_\alpha^{\circ 4}$	$A_\alpha^{\circ 5}$	$A_\alpha^{\circ 6}$	$A_\alpha^{\circ 7}$	$A_\alpha^{\circ 8}$	$A_\alpha^{\circ 9}$	$A_\alpha^{\circ 10}$	$A_\alpha^{\circ 11}$	$A_\alpha^{\circ 12}$	$A_\alpha^{\circ 13}$	$A_\alpha^{\circ 0}$	$A_\alpha^{\circ 1}$
$A_\alpha^{\circ 3}$	$A_\alpha^{\circ 3}$	$A_\alpha^{\circ 4}$	$A_\alpha^{\circ 5}$	$A_\alpha^{\circ 6}$	$A_\alpha^{\circ 7}$	$A_\alpha^{\circ 8}$	$A_\alpha^{\circ 9}$	$A_\alpha^{\circ 10}$	$A_\alpha^{\circ 11}$	$A_\alpha^{\circ 12}$	$A_\alpha^{\circ 13}$	$A_\alpha^{\circ 0}$	$A_\alpha^{\circ 1}$	$A_\alpha^{\circ 2}$
$A_\alpha^{\circ 4}$	$A_\alpha^{\circ 4}$	$A_\alpha^{\circ 5}$	$A_\alpha^{\circ 6}$	$A_\alpha^{\circ 7}$	$A_\alpha^{\circ 8}$	$A_\alpha^{\circ 9}$	$A_\alpha^{\circ 10}$	$A_\alpha^{\circ 11}$	$A_\alpha^{\circ 12}$	$A_\alpha^{\circ 13}$	$A_\alpha^{\circ 0}$	$A_\alpha^{\circ 1}$	$A_\alpha^{\circ 2}$	$A_\alpha^{\circ 3}$
$A_\alpha^{\circ 5}$	$A_\alpha^{\circ 5}$	$A_\alpha^{\circ 6}$	$A_\alpha^{\circ 7}$	$A_\alpha^{\circ 8}$	$A_\alpha^{\circ 9}$	$A_\alpha^{\circ 10}$	$A_\alpha^{\circ 11}$	$A_\alpha^{\circ 12}$	$A_\alpha^{\circ 13}$	$A_\alpha^{\circ 0}$	$A_\alpha^{\circ 1}$	$A_\alpha^{\circ 2}$	$A_\alpha^{\circ 3}$	$A_\alpha^{\circ 4}$
$A_\alpha^{\circ 6}$	$A_\alpha^{\circ 6}$	$A_\alpha^{\circ 7}$	$A_\alpha^{\circ 8}$	$A_\alpha^{\circ 9}$	$A_\alpha^{\circ 10}$	$A_\alpha^{\circ 11}$	$A_\alpha^{\circ 12}$	$A_\alpha^{\circ 13}$	$A_\alpha^{\circ 0}$	$A_\alpha^{\circ 1}$	$A_\alpha^{\circ 2}$	$A_\alpha^{\circ 3}$	$A_\alpha^{\circ 4}$	$A_\alpha^{\circ 5}$
$A_\alpha^{\circ 7}$	$A_\alpha^{\circ 7}$	$A_\alpha^{\circ 8}$	$A_\alpha^{\circ 9}$	$A_\alpha^{\circ 10}$	$A_\alpha^{\circ 11}$	$A_\alpha^{\circ 12}$	$A_\alpha^{\circ 13}$	$A_\alpha^{\circ 0}$	$A_\alpha^{\circ 1}$	$A_\alpha^{\circ 2}$	$A_\alpha^{\circ 3}$	$A_\alpha^{\circ 4}$	$A_\alpha^{\circ 5}$	$A_\alpha^{\circ 6}$
$A_\alpha^{\circ 8}$	$A_\alpha^{\circ 8}$	$A_\alpha^{\circ 9}$	$A_\alpha^{\circ 10}$	$A_\alpha^{\circ 11}$	$A_\alpha^{\circ 12}$	$A_\alpha^{\circ 13}$	$A_\alpha^{\circ 0}$	$A_\alpha^{\circ 1}$	$A_\alpha^{\circ 2}$	$A_\alpha^{\circ 3}$	$A_\alpha^{\circ 4}$	$A_\alpha^{\circ 5}$	$A_\alpha^{\circ 6}$	$A_\alpha^{\circ 7}$
$A_\alpha^{\circ 9}$	$A_\alpha^{\circ 9}$	$A_\alpha^{\circ 10}$	$A_\alpha^{\circ 11}$	$A_\alpha^{\circ 12}$	$A_\alpha^{\circ 13}$	$A_\alpha^{\circ 0}$	$A_\alpha^{\circ 1}$	$A_\alpha^{\circ 2}$	$A_\alpha^{\circ 3}$	$A_\alpha^{\circ 4}$	$A_\alpha^{\circ 5}$	$A_\alpha^{\circ 6}$	$A_\alpha^{\circ 7}$	$A_\alpha^{\circ 8}$
$A_\alpha^{\circ 10}$	$A_\alpha^{\circ 10}$	$A_\alpha^{\circ 11}$	$A_\alpha^{\circ 12}$	$A_\alpha^{\circ 13}$	$A_\alpha^{\circ 0}$	$A_\alpha^{\circ 1}$	$A_\alpha^{\circ 2}$	$A_\alpha^{\circ 3}$	$A_\alpha^{\circ 4}$	$A_\alpha^{\circ 5}$	$A_\alpha^{\circ 6}$	$A_\alpha^{\circ 7}$	$A_\alpha^{\circ 8}$	$A_\alpha^{\circ 9}$
$A_\alpha^{\circ 11}$	$A_\alpha^{\circ 11}$	$A_\alpha^{\circ 12}$	$A_\alpha^{\circ 13}$	$A_\alpha^{\circ 0}$	$A_\alpha^{\circ 1}$	$A_\alpha^{\circ 2}$	$A_\alpha^{\circ 3}$	$A_\alpha^{\circ 4}$	$A_\alpha^{\circ 5}$	$A_\alpha^{\circ 6}$	$A_\alpha^{\circ 7}$	$A_\alpha^{\circ 8}$	$A_\alpha^{\circ 9}$	$A_\alpha^{\circ 10}$
$A_\alpha^{\circ 12}$	$A_\alpha^{\circ 12}$	$A_\alpha^{\circ 13}$	$A_\alpha^{\circ 0}$	$A_\alpha^{\circ 1}$	$A_\alpha^{\circ 2}$	$A_\alpha^{\circ 3}$	$A_\alpha^{\circ 4}$	$A_\alpha^{\circ 5}$	$A_\alpha^{\circ 6}$	$A_\alpha^{\circ 7}$	$A_\alpha^{\circ 8}$	$A_\alpha^{\circ 9}$	$A_\alpha^{\circ 10}$	$A_\alpha^{\circ 11}$
$A_\alpha^{\circ 13}$	$A_\alpha^{\circ 13}$	$A_\alpha^{\circ 0}$	$A_\alpha^{\circ 1}$	$A_\alpha^{\circ 2}$	$A_\alpha^{\circ 3}$	$A_\alpha^{\circ 4}$	$A_\alpha^{\circ 5}$	$A_\alpha^{\circ 6}$	$A_\alpha^{\circ 7}$	$A_\alpha^{\circ 8}$	$A_\alpha^{\circ 9}$	$A_\alpha^{\circ 10}$	$A_\alpha^{\circ 11}$	$A_\alpha^{\circ 12}$

Corollary 1.4 *Let $h : \mathbb{R}^m \to \mathbb{R}^m$ be a function such that $\exists_m\mathrm{MO}_{x,\alpha}^{\infty,u}(h)$. So, if it is fulfilled the following condition*

$$ \forall o_x^\alpha \in {}_m\mathrm{MO}_{x,\alpha}^{\infty,u}(h) \quad \exists_m G\big(A_\alpha(o_x^\alpha)\big) \subset {}_mG_{FIM}(\alpha), \tag{33} $$

such that ${}_mG\big(A_\alpha(o_x^\alpha)\big)$ is the group generated by the operator $A_\alpha(o_x^\alpha)$. As a consequence, it is fulfilled that

$$ {}_mG_{FIM}(\alpha) = \bigcup_{o_x^\alpha \in {}_m\mathrm{MO}_{x,\alpha}^{\infty,u}(h)_m} G\big(A_\alpha(o_x^\alpha)\big). \tag{34} $$

It is necessary to mention that the Corollary 1.3 allows generating groups of fractional operators under other operations, as shown in the following corollary:

Corollary 1.5 *Let $\mathbb{Z}_p^+$ be the set of positive residual classes less than p, with p a prime number. So, for each fractional operator $o_x^\alpha \in {}_m\mathrm{MO}_{x,\alpha}^{\infty,u}(h)$, it is possible to define the following set of fractional matrix operators*

$$ {}_mG^*\left(A_\alpha\left(o_x^\alpha\right), \mathbb{Z}_p^+\right) := \left\{A_\alpha^{\circ r} = A_\alpha\left(o_x^{r\alpha}\right) \ : \ r \in \mathbb{Z}_p^+ \quad \text{and} \quad A_\alpha^{\circ r} = \left(\left[A_\alpha^{\circ r}\right]_{jk}\right) := \left(o_k^{r\alpha}\right)\right\}, \tag{35} $$

which corresponds to an Abelian group under the following operation

$$ A_\alpha^{\circ r} * A_\alpha^{\circ s} = A_\alpha^{\circ rs}. \tag{36} $$

Example 2 *Let o_x^α be a fractional operator such that $o_x^\alpha \in {}_mMO_{x,\alpha}^{\infty,u}(h)$. So, considering the set $\mathbb{Z}_{13}^+$, it is possible to define the Abelian group of fractional matrix operators ${}_mG^*\left(A_\alpha\left(o_x^\alpha\right), \mathbb{Z}_{13}^+\right)$. Furthermore, all possible combinations of the elements of the group under the operation (36) are summarized below:*

$*$	$A_\alpha^{\circ 1}$	$A_\alpha^{\circ 2}$	$A_\alpha^{\circ 3}$	$A_\alpha^{\circ 4}$	$A_\alpha^{\circ 5}$	$A_\alpha^{\circ 6}$	$A_\alpha^{\circ 7}$	$A_\alpha^{\circ 8}$	$A_\alpha^{\circ 9}$	$A_\alpha^{\circ 10}$	$A_\alpha^{\circ 11}$	$A_\alpha^{\circ 12}$
$A_\alpha^{\circ 1}$	$A_\alpha^{\circ 1}$	$A_\alpha^{\circ 2}$	$A_\alpha^{\circ 3}$	$A_\alpha^{\circ 4}$	$A_\alpha^{\circ 5}$	$A_\alpha^{\circ 6}$	$A_\alpha^{\circ 7}$	$A_\alpha^{\circ 8}$	$A_\alpha^{\circ 9}$	$A_\alpha^{\circ 10}$	$A_\alpha^{\circ 11}$	$A_\alpha^{\circ 12}$
$A_\alpha^{\circ 2}$	$A_\alpha^{\circ 2}$	$A_\alpha^{\circ 4}$	$A_\alpha^{\circ 6}$	$A_\alpha^{\circ 8}$	$A_\alpha^{\circ 10}$	$A_\alpha^{\circ 12}$	$A_\alpha^{\circ 1}$	$A_\alpha^{\circ 3}$	$A_\alpha^{\circ 5}$	$A_\alpha^{\circ 7}$	$A_\alpha^{\circ 9}$	$A_\alpha^{\circ 11}$
$A_\alpha^{\circ 3}$	$A_\alpha^{\circ 3}$	$A_\alpha^{\circ 6}$	$A_\alpha^{\circ 9}$	$A_\alpha^{\circ 12}$	$A_\alpha^{\circ 2}$	$A_\alpha^{\circ 5}$	$A_\alpha^{\circ 8}$	$A_\alpha^{\circ 11}$	$A_\alpha^{\circ 1}$	$A_\alpha^{\circ 4}$	$A_\alpha^{\circ 7}$	$A_\alpha^{\circ 10}$
$A_\alpha^{\circ 4}$	$A_\alpha^{\circ 4}$	$A_\alpha^{\circ 8}$	$A_\alpha^{\circ 12}$	$A_\alpha^{\circ 3}$	$A_\alpha^{\circ 7}$	$A_\alpha^{\circ 11}$	$A_\alpha^{\circ 2}$	$A_\alpha^{\circ 6}$	$A_\alpha^{\circ 10}$	$A_\alpha^{\circ 1}$	$A_\alpha^{\circ 5}$	$A_\alpha^{\circ 9}$
$A_\alpha^{\circ 5}$	$A_\alpha^{\circ 5}$	$A_\alpha^{\circ 10}$	$A_\alpha^{\circ 2}$	$A_\alpha^{\circ 7}$	$A_\alpha^{\circ 12}$	$A_\alpha^{\circ 4}$	$A_\alpha^{\circ 9}$	$A_\alpha^{\circ 1}$	$A_\alpha^{\circ 6}$	$A_\alpha^{\circ 11}$	$A_\alpha^{\circ 3}$	$A_\alpha^{\circ 8}$
$A_\alpha^{\circ 6}$	$A_\alpha^{\circ 6}$	$A_\alpha^{\circ 12}$	$A_\alpha^{\circ 5}$	$A_\alpha^{\circ 11}$	$A_\alpha^{\circ 4}$	$A_\alpha^{\circ 10}$	$A_\alpha^{\circ 3}$	$A_\alpha^{\circ 9}$	$A_\alpha^{\circ 2}$	$A_\alpha^{\circ 8}$	$A_\alpha^{\circ 1}$	$A_\alpha^{\circ 7}$
$A_\alpha^{\circ 7}$	$A_\alpha^{\circ 7}$	$A_\alpha^{\circ 1}$	$A_\alpha^{\circ 8}$	$A_\alpha^{\circ 2}$	$A_\alpha^{\circ 9}$	$A_\alpha^{\circ 3}$	$A_\alpha^{\circ 10}$	$A_\alpha^{\circ 4}$	$A_\alpha^{\circ 11}$	$A_\alpha^{\circ 5}$	$A_\alpha^{\circ 12}$	$A_\alpha^{\circ 6}$
$A_\alpha^{\circ 8}$	$A_\alpha^{\circ 8}$	$A_\alpha^{\circ 3}$	$A_\alpha^{\circ 11}$	$A_\alpha^{\circ 6}$	$A_\alpha^{\circ 1}$	$A_\alpha^{\circ 9}$	$A_\alpha^{\circ 4}$	$A_\alpha^{\circ 12}$	$A_\alpha^{\circ 7}$	$A_\alpha^{\circ 2}$	$A_\alpha^{\circ 10}$	$A_\alpha^{\circ 5}$
$A_\alpha^{\circ 9}$	$A_\alpha^{\circ 9}$	$A_\alpha^{\circ 5}$	$A_\alpha^{\circ 1}$	$A_\alpha^{\circ 10}$	$A_\alpha^{\circ 6}$	$A_\alpha^{\circ 2}$	$A_\alpha^{\circ 11}$	$A_\alpha^{\circ 7}$	$A_\alpha^{\circ 3}$	$A_\alpha^{\circ 12}$	$A_\alpha^{\circ 8}$	$A_\alpha^{\circ 4}$
$A_\alpha^{\circ 10}$	$A_\alpha^{\circ 10}$	$A_\alpha^{\circ 7}$	$A_\alpha^{\circ 4}$	$A_\alpha^{\circ 1}$	$A_\alpha^{\circ 11}$	$A_\alpha^{\circ 8}$	$A_\alpha^{\circ 5}$	$A_\alpha^{\circ 2}$	$A_\alpha^{\circ 12}$	$A_\alpha^{\circ 9}$	$A_\alpha^{\circ 6}$	$A_\alpha^{\circ 3}$
$A_\alpha^{\circ 11}$	$A_\alpha^{\circ 11}$	$A_\alpha^{\circ 9}$	$A_\alpha^{\circ 7}$	$A_\alpha^{\circ 5}$	$A_\alpha^{\circ 3}$	$A_\alpha^{\circ 1}$	$A_\alpha^{\circ 12}$	$A_\alpha^{\circ 10}$	$A_\alpha^{\circ 8}$	$A_\alpha^{\circ 6}$	$A_\alpha^{\circ 4}$	$A_\alpha^{\circ 2}$
$A_\alpha^{\circ 12}$	$A_\alpha^{\circ 12}$	$A_\alpha^{\circ 11}$	$A_\alpha^{\circ 10}$	$A_\alpha^{\circ 9}$	$A_\alpha^{\circ 8}$	$A_\alpha^{\circ 7}$	$A_\alpha^{\circ 6}$	$A_\alpha^{\circ 5}$	$A_\alpha^{\circ 4}$	$A_\alpha^{\circ 3}$	$A_\alpha^{\circ 2}$	$A_\alpha^{\circ 1}$

On the other hand, defining $A_\alpha(h) = \left(\left[A_\alpha(h)\right]_{jk}\right) := \left([h]_k\right)$, it is possible to obtain the following result:

$$ \forall A_\alpha^{\circ r} \in {}_mG_{FIM}(\alpha) \quad \exists A_{h,r\alpha} \in {}_m\mathrm{M}_{x,\alpha}^{\infty}(h) \quad \text{such that} \quad A_{h,r\alpha} := A_\alpha\left(o_x^{r\alpha}\right) \circ A_\alpha^T(h). \tag{37} $$

Therefore, if Φ_{FIM} denotes the iteration function of some fractional iterative method [26], it is possible to obtain the following results:

$$ \begin{aligned} \text{Let } \alpha_0 \in \mathbb{R}\backslash\mathbb{Z} \Rightarrow \ & \forall A_{h,\alpha_0} \in {}_m\mathrm{M}_{x,\alpha}^{\infty}(h) \exists \Phi_{FIM} \\ & = \Phi_{FIM}(A_{h,\alpha_0}) .\because. \forall A_{h,\alpha_0} \quad \exists\{\Phi_{FIM}(A_{h,\alpha}) : \ \alpha \in \mathbb{R}\backslash\mathbb{Z}\}, \end{aligned} \tag{38} $$

$$\begin{aligned}\text{Let }\ \alpha_0\in\mathbb{R}\backslash\mathbb{Z}\Rightarrow\ &\forall A_{\alpha_0}^{\circ 1}\in{}_m\mathrm{G}_{FIM}(\alpha)\exists\Phi_{FIM}\\&=\Phi_{FIM}(A_{\alpha_0})\therefore\forall A_{\alpha_0}\quad\exists\{\Phi_{FIM}(A_\alpha):\ \alpha\in\mathbb{R}\backslash\mathbb{Z}\}.\end{aligned}\tag{39}$$

To finish this section, it is necessary to mention that the applications of fractional operators have spread to different fields of science such as finance [27, 28], economics [29], number theory through the Riemann zeta function [30, 31], and in engineering with the study for the manufacture of hybrid solar receivers [32, 33]. It is worth mentioning that there exists also a growing interest in fractional operators and their properties for solving nonlinear algebraic systems [24, 34–41], which is a classical problem in mathematics, physics and engineering, which consists of finding the set of zeros of a function $f:\Omega\subset\mathbb{R}^n\to\mathbb{R}^n$, that is,

$$\{\xi\in\Omega\ :\ \|f(\xi)\|=0\},\tag{40}$$

where $\|\cdot\|:\mathbb{R}^n\to\mathbb{R}$ denotes any vector norm, or equivalently

$$\{\xi\in\Omega\ :\ [f]_k(\xi)=0\quad\forall k\geq 1\}.\tag{41}$$

Although finding the zeros of a function may seem like a simple problem, it is generally necessary to use numerical methods of the iterative type to solve it.

4. Fixed-point method

Let $\Phi:\mathbb{R}^n\to\mathbb{R}^n$ be a function. It is possible to build a sequence $\{x_i\}_{i\geq1}$ by defining the following iterative method

$$x_{i+1}:=\Phi(x_i),\quad i=0,1,2,\cdots.\tag{42}$$

So, if it is fulfilled that $x_i\to\xi\in\mathbb{R}^n$ and the function Φ is continuous around ξ, we obtain that

$$\xi=\lim_{i\to\infty}x_{i+1}=\lim_{i\to\infty}\Phi(x_i)=\Phi\left(\lim_{i\to\infty}x_i\right)=\Phi(\xi),\tag{43}$$

the above result is the reason by which the method (42) is known as the fixed-point method. Furthermore, the function Φ is called an iteration function. On the other hand, considering the following set

$$B(\xi;\delta):=\{x\ :\ \|x-\xi\|<\delta\},\tag{44}$$

it is possible to define the following corollary, which allows characterizing the order of convergence of an iteration function Φ through its Jacobian matrix $\Phi^{(1)}$ [7]:

Corollary 1.6 *Let $\Phi:\mathbb{R}^n\to\mathbb{R}^n$ be an iteration function. If Φ defines a sequence $\{x_i\}_{i\geq1}$ such that $x_i\to\xi\in\mathbb{R}^n$. So, Φ has an order of convergence of order (at least) p in $B(\xi;\delta)$, where it is fulfilled that:*

$$p:=\begin{cases}1, & \text{if }\ \lim_{x\to\xi}\left\|\Phi^{(1)}(x)\right\|\neq0\\ 2, & \text{if }\ \lim_{x\to\xi}\left\|\Phi^{(1)}(x)\right\|=0\end{cases}.\tag{45}$$

5. Fractional fixed-point method

Let $\mathbb{N}_0$ be the set $\mathbb{N}\cup\{0\}$, if $\gamma\in\mathbb{N}_0^m$ and $x\in\mathbb{R}^m$, then it is possible to define the following multi-index notation

$$\begin{cases} \gamma! := \prod_{k=1}^{m} [\gamma]_k!, \quad |\gamma| := \sum_{k=1}^{m} [\gamma]_k 0.1cm, \quad x^\gamma := \prod_{k=1}^{m} [x]_k^{[\gamma]_k} \\ \dfrac{\partial^\gamma}{\partial x^\gamma} := \dfrac{\partial^{[\gamma]_1}}{\partial [x]_1^{[\gamma]_1}} \dfrac{\partial^{[\gamma]_2}}{\partial [x]_2^{[\gamma]_2}} \cdots \dfrac{\partial^{[\gamma]_m}}{\partial [x]_m^{[\gamma]_m}} \end{cases}. \tag{46}$$

So, considering a function $h : \Omega\subset\mathbb{R}^m \to \mathbb{R}$ and the fractional operator

$$s_x^{\alpha\gamma}\left(o_x^\alpha\right) := o_1^{\alpha[\gamma]_1} o_2^{\alpha[\gamma]_2} \cdots o_m^{\alpha[\gamma]_m}, \tag{47}$$

it is possible to define the following set of fractional operators

$$S_{x,\alpha}^{n,\gamma}(h) := \left\{ s_x^{\alpha\gamma} = s_x^{\alpha\gamma}\left(o_x^\alpha\right) : \exists s_x^{\alpha\gamma} h(x) \text{ with } o_x^\alpha \in O_{x,\alpha}^s(h)\ \forall s\le n^2 \right.$$
$$\left. \text{and } \lim_{\alpha\to k} s_x^{\alpha\gamma} h(x) = \frac{\partial^{k\gamma}}{\partial x^{k\gamma}} h(x)\ \forall \alpha, |\gamma|\le n \right\}, \tag{48}$$

from which it is possible to obtain the following results:

$$\text{If } s_x^{\alpha\gamma}\in S_{x,\alpha}^{n,\gamma}(h) \Rightarrow \begin{cases} \lim_{\alpha\to 0} s_x^{\alpha\gamma} h(x) = o_1^0 o_2^0 \cdots o_m^0 h(x) = h(x) \\ \lim_{\alpha\to 1} s_x^{\alpha\gamma} h(x) = o_1^{[\gamma]_1} o_2^{[\gamma]_2} \cdots o_m^{[\gamma]_m} h(x) = \dfrac{\partial^\gamma}{\partial x^\gamma} h(x)\ \forall |\gamma|\le n \\ \lim_{\alpha\to q} s_x^{\alpha\gamma} h(x) = o_1^{q[\gamma]_1} o_2^{q[\gamma]_2} \cdots o_m^{q[\gamma]_m} h(x) = \dfrac{\partial^{q\gamma}}{\partial x^{q\gamma}} h(x)\ \forall q|\gamma|\le qn \\ \lim_{\alpha\to n} s_x^{\alpha\gamma} h(x) = o_1^{n[\gamma]_1} o_2^{n[\gamma]_2} \cdots o_m^{n[\gamma]_m} h(x) = \dfrac{\partial^{n\gamma}}{\partial x^{n\gamma}} h(x)\ \forall n|\gamma|\le n^2 \end{cases}, \tag{49}$$

and as a consequence, considering a function $h : \Omega\subset\mathbb{R}^m \to \mathbb{R}^m$, it is possible to define the following set of fractional operators

$$_m S_{x,\alpha}^{n,\gamma}(h) := \left\{ s_x^{\alpha\gamma} \ :\ s_x^{\alpha\gamma} \in S_{x,\alpha}^{n,\gamma}\left([h]_k\right)\ \forall k\le m \right\}. \tag{50}$$

On the other hand, using little-o notation it is possible to obtain the following result:

$$\text{If } x\in B(a;\delta) \Rightarrow \lim_{x\to a} \frac{o((x-a)^\gamma)}{(x-a)^\gamma} \to 0 \quad \forall |\gamma|\ge 1, \tag{51}$$

with which it is possible to define the following set of functions

$$R_{\alpha\gamma}^n(a) := \left\{ r_{\alpha\gamma}^n \ :\ \lim_{x\to a} \left\| r_{\alpha\gamma}^n(x) \right\| = 0\ \forall |\gamma|\ge n \text{ and } \left\| r_{\alpha\gamma}^n(x) \right\| \le o(\|x-a\|^n)\ \forall x\in B(a;\delta) \right\}, \tag{52}$$

where $r^n_{\alpha\gamma} : B(a;\delta) \subset \Omega \to \mathbb{R}^m$. So, considering the previous set and some $B(a;\delta) \subset \Omega$, it is possible to define the following sets of fractional operators

$$mT^{n,q,\gamma}_{x,\alpha,p}(a,h) := \left\{ t^{\alpha,p}_x = t^{\alpha,p}_x\left(s^{\alpha\gamma}_x\right) : \ s^{\alpha\gamma}_x \in {}_mS^{M,\gamma}_{x,\alpha}(h) \quad \text{and} \right.$$

$$\left. t^{\alpha,p}_x h(x) := \sum_{|\gamma|=0}^{p} \frac{1}{\gamma!} \hat{e}_j s^{\alpha\gamma}_x[h]_j(a)(x-a)^\gamma + r^p_{\alpha\gamma}(x) \begin{array}{c} \forall \alpha \le n \\ \forall p \le q \end{array} \right\}, \tag{53}$$

$$mT^{\infty,\gamma}_{x,\alpha}(a,h) := \left\{ t^{\alpha,\infty}_x = t^{\alpha,\infty}_x\left(s^{\alpha\gamma}_x\right) : \ s^{\alpha\gamma}_x \in {}_mS^{\infty,\gamma}_{x,\alpha}(h) \quad \text{and} \right.$$

$$\left. t^{\alpha,\infty}_x h(x) := \sum_{|\gamma|=0}^{\infty} \frac{1}{\gamma!} \hat{e}_j s^{\alpha\gamma}_x[h]_j(a)(x-a)^\gamma \right\}, \tag{54}$$

which allow generalizing the Taylor series expansion of a vector-valued function in multi-index notation [7], where $M = \max\{n, q\}$. As a consequence, it is possible to obtain the following results:

$$\text{If } \ t^{\alpha,p}_x \in {}_mT^{1,q,\gamma}_{x,\alpha,p}(a,h) \ \text{ and } \ \alpha \to 1 \ \Rightarrow$$

$$t^{1,p}_x h(x) = h(a) + \sum_{|\gamma|=1}^{p} \frac{1}{\gamma!} \hat{e}_j \frac{\partial^\gamma}{\partial x^\gamma}[h]_j(a)(x-a)^\gamma + r^p_\gamma(x), \tag{55}$$

$$\text{If } \ t^{\alpha,p}_x \in {}_mT^{n,1,\gamma}_{x,\alpha,p}(a,h) \ \text{ and } \ p \to 1 \ \Rightarrow \ t^{\alpha,1}_x h(x) = h(a) + \sum_{k=1}^{m} \hat{e}_j o^\alpha_k[h]_j(a)[(x-a)]_k + r^1_{\alpha\gamma}(x). \tag{56}$$

Let $f : \Omega \subset \mathbb{R}^n \to \mathbb{R}^n$ be a function with a point $\xi \in \Omega$ such that $\|f(\xi)\| = 0$. So, for some $x_i \in B(\xi;\delta) \subset \Omega$ and for some fractional operator $t^{\alpha,\infty}_x \in {}_nT^{\infty,\gamma}_{x,\alpha}(x_i, f)$, it is possible to define a type of linear approximation of the function f around the value x_i as follows

$$t^{\alpha,\infty}_x f(x) \approx f(x_i) + \sum_{k=1}^{n} \hat{e}_j o^\alpha_k[f]_j(x_i)[(x-x_i)]_k, \tag{57}$$

which may be rewritten more compactly as follows

$$t^{\alpha,\infty}_x f(x) \approx f(x_i) + \left(o^\alpha_k[f]_j(x_i)\right)(x - x_i). \tag{58}$$

where $\left(o^\alpha_k[f]_j(x_i)\right)$ denotes a square matrix. On the other hand, if $x \to \xi$ and since $\|f(\xi)\| = 0$, it follows that

$$0 \approx f(x_i) + \left(o^\alpha_k[f]_j(x_i)\right)(\xi - x_i) \quad \Rightarrow \quad \xi \approx x_i - \left(o^\alpha_k[f]_j(x_i)\right)^{-1} f(x_i). \tag{59}$$

So, defining the following matrix

$$A_{f,\alpha}(x) = \left([A_{f,\alpha}]_{jk}(x)\right) := \left(o^\alpha_k[f]_j(x)\right)^{-1}, \tag{60}$$

it is possible to define the following fractional iterative method

$$x_{i+1} := \Phi(\alpha, x_i) = x_i - A_{f,\alpha}(x_i) f(x_i), \quad i = 0,1,2,\cdots, \tag{61}$$

which corresponds to the more general case of the fractional Newton-Raphson method [25, 36].

Let $f : \Omega \subset \mathbb{R}^n \to \mathbb{R}^n$ be a function with a point $\xi \in \Omega$ such that $\|f(\xi)\| = 0$. So, considering an iteration function $\Phi : (\mathbb{R}\backslash\mathbb{Z}) \times \mathbb{R}^n \to \mathbb{R}^n$, the iteration function of a fractional iterative method may be written in general form as follows

$$\Phi(\alpha, x) := x - A_{g,\alpha}(x) f(x), \quad \alpha \in \mathbb{R}\backslash\mathbb{Z}, \tag{62}$$

where $A_{g,\alpha}$ is a matrix that depends, in at least one of its entries, on fractional operators of order α applied to some function $g : \mathbb{R}^n \to \mathbb{R}^n$, whose particular case occurs when $g = f$. So, it is possible to define in a general way a fractional fixed-point method as follows

$$x_{i+1} := \Phi(\alpha, x_i), \quad i = 0,1,2,\cdots. \tag{63}$$

Before continuing, it is worth mentioning that one of the main advantages of fractional iterative methods is that the initial condition x_0 can remain fixed, with which it is enough to vary the order α of the fractional operators involved until generating a sequence convergent $\{x_i\}_{i\geq 1}$ to the value $\xi \in \Omega$. Since the order α of the fractional operators is varied, different values of α can generate different convergent sequences to the same value ξ but with a different number of iterations. So, it is possible to define the following set

$$\text{Conv}_\delta(\xi) := \left\{ \Phi : \lim_{x\to\xi} \Phi(\alpha, \ x) = \xi_\alpha \in B(\xi; \delta) \right\} \tag{64}$$

which may be interpreted as the set of fractional fixed-point methods that define a convergent sequence $\{x_i\}_{i\geq 1}$ to some value $\xi_\alpha \in B(\xi; \delta)$. So, denoting by $\text{card}(\cdot)$ the cardinality of a set, under certain conditions it is possible to prove the following result (see reference [24], proof of Theorem 2):

$$\text{card}(\text{Conv}_\delta(\xi)) = \text{card}(\mathbb{R}), \tag{65}$$

from which it follows that the set (64) is generated by an uncountable family of fractional fixed-point methods. Before continuing, it is necessary to define the following proposition [7]:

Proposition 1.7 *Let* $\Phi : (\mathbb{R}\backslash\mathbb{Z}) \times \mathbb{R}^n \to \mathbb{R}^n$ *be an iteration function such that* $\Phi \in \text{Conv}_\delta(\xi)$ *in a region* Ω. *So, if* Φ *is given by the equation (62) and fulfills the following condition*

$$\lim_{x\to\xi} A_{g,\alpha}(x) = \left(f^{(1)}(\xi)\right)^{-1}. \tag{66}$$

Then, Φ fulfills a necessary (but not sufficient) condition to be convergent of order (at least) quadratic in $B(\xi; \delta)$.

Proof: If Φ is given by the equation (62), the k-th component of the function Φ may be written as follows

$$[\Phi]_k(\alpha, x) = [x]_k - \sum_{j=1}^{n} [A_{g,\alpha}]kj(x)[f]_j(x), \tag{67}$$

and considering that $f^{(1)}(x) = \left(\left[f^{(1)}\right]jl(x)\right) := (\partial l[f]j(x))$, it is possible to obtain the following result

$$\left[\Phi^{(1)}\right]_{kl}(\alpha, x) = \partial_l[\Phi]_k(\alpha, x) = \delta_{kl} - \sum_{j=1}^{n}\left(\left[A_{g,\alpha}\right]_{kj}(x)\left[f^{(1)}\right]_{jl}(x) + \left(\partial_l\left[A_{g,\alpha}\right]_{kj}(x)\right)[f]_j(x)\right), \tag{68}$$

where δ_{kl} denotes the Kronecker delta. On the other hand, since f has a point $\xi \in \Omega$ such that $\|f(\xi)\| = 0$, it follows that

$$\left[\Phi^{(1)}\right]_{kl}(\alpha, \xi) = \delta_{kl} - \sum_{j=1}^{n}\left[A_{g,\alpha}\right]_{kj}(\xi)\left[f^{(1)}\right]_{jl}(\xi). \tag{69}$$

Then, if $\Phi \in \mathrm{Conv}_\delta(\xi)$ and has an order of convergence (at least) quadratic in $B(\xi; \delta)$, by the Corollary 1.6, it is fulfilled the following condition

$$\sum_{j=1}^{n}\left[A_{g,a}\right]_{kj}(\xi)\left[f^{(1)}\right]_{jl}(\xi) = \delta_{kl}, \quad \forall k,l \leq n, \tag{70}$$

which may be rewritten more compactly as follows

$$A_{g,a}(\xi)f^{(1)}(\xi) = I_n, \tag{71}$$

where I_n denotes the identity matrix of $n \times n$. Therefore, any matrix $A_{g,a}$ that fulfills the following condition

$$\lim_{x \to \xi} A_{g,a}(x) = \left(f^{(1)}(\xi)\right)^{-1}, \tag{72}$$

ensures that the iteration function Φ given by the equation (62), fulfills a necessary (but not sufficient) condition to be convergent of order (at least) quadratic in $B(\xi; \delta)$.

Considering the Corollary 1.6 and the Proposition 1.7, it is possible to define the following sets to classify the order of convergence of some fractional iterative methods:

$$\mathrm{Ord}^1(\xi) := \left\{\Phi \in \mathrm{Conv}_\delta(\xi) : \lim_{x \to \xi}\left\|\Phi^{(1)}(a, \ x)\right\| \neq 0\right\}, \tag{73}$$

$$\mathrm{Ord}^2(\xi) := \left\{\Phi \in \mathrm{Conv}_\delta(\xi) : \lim_{x \to \xi}\left\|\Phi^{(1)}(a, \ x)\right\| = 0\right\}, \tag{74}$$

$$\mathrm{ord}^1(\xi) := \left\{\Phi \in \mathrm{Conv}_\delta(\xi) : \lim_{x \to \xi} A_{g,a}(x) \neq \left(f^{(1)}(\xi)\right)^{-1} \text{ or } \lim_{\alpha \to 1} A_{g,a}(\xi) \neq \left(f^{(1)}(\xi)\right)^{-1}\right\}, \tag{75}$$

$$\mathrm{ord}^2(\xi) := \left\{\Phi \in \mathrm{Conv}_\delta(\xi) : \lim_{x \to \xi} A_{g,a}(x) \neq \left(f^{(1)}(\xi)\right)^{-1} \text{ or } \lim_{\alpha \to 1} A_{g,a}(\xi) \neq \left(f^{(1)}(\xi)\right)^{-1}\right\}, \tag{76}$$

On the other hand, considering that depending on the nature of the function f, there exist cases in which the Newton-Raphson method can present an order of convergence (at least) linear [7]. So, it is possible to obtain the following relations between the previous sets

$$\mathrm{ord}^1(\xi) \subset \mathrm{Ord}^1(\xi) \quad \text{and} \quad \mathrm{ord}^2(\xi) \subset \mathrm{Ord}^1(\xi) \cup \mathrm{Ord}^2(\xi), \tag{77}$$

with which it is possible to define the following sets

$$\mathrm{Ord}_2^1(\xi) := \mathrm{ord}^2(\xi) \cap \mathrm{Ord}^1(\xi) \quad \text{and} \quad \mathrm{Ord}_2^2(\xi) := \mathrm{ord}^2(\xi) \cap \mathrm{Ord}^2(\xi). \tag{78}$$

5.1 Acceleration of the order of convergence of the set $\mathrm{Ord}_2^1(\xi)$

Let $f : \Omega \subset \mathbb{R}^n \to \mathbb{R}^n$ be a function with a point $\xi \in \Omega$ such that $\|f(\xi)\| = 0$, and denoting by Φ_{NR} to the iteration function of the Newton-Raphson method, it is possible to define the following set of functions

$$\mathrm{Ord}_{NR}^2(\xi) := \left\{ f : \lim_{x \to \xi} \left\| \Phi_{NR}^{(1)}(x) \right\| = 0 \right\}. \tag{79}$$

So, it is possible to define the following corollary:

Corollary 1.8 *Let $f : \Omega \subset \mathbb{R}^n \to \mathbb{R}^n$ be a function such that $f \in \mathrm{Ord}_{NR}^2(\xi)$, and let $\Phi : (\mathbb{R} \backslash \mathbb{Z}) \times \mathbb{R}^n \to \mathbb{R}^n$ be an iteration function given by the equation (62) such that $\Phi \in \mathrm{ord}^1(\xi)$. So, if Φ also fulfills the following condition*

$$\lim_{\alpha \to 1} A_{g,\alpha}(\xi) = \left(f^{(1)}(\xi) \right)^{-1}. \tag{80}$$

Then, $\Phi \in \mathrm{Ord}_2^1(\xi)$. Therefore, it is possible to assign a positive value δ_0, and replace the order α of the fractional operators of the matrix $A_{g,\alpha}$ by the following function

$$\alpha_f\left([x]_k, x\right) := \begin{cases} \alpha, & \text{if} |[x]_k| \neq 0 \quad \text{and} \quad \|f(x)\| \geq \delta_0 \\ 1 & \text{if} |[x]_k| = 0 \quad \text{or} \quad \|f(x)\| \geq \delta_0 \end{cases}, \tag{81}$$

obtaining a new matrix that may be denoted as follows

$$A_{g,\alpha_f}(x) = \left(\left[A_{g,\alpha_f} \right]_{jk}(x) \right), \quad \alpha \in \mathbb{R} \backslash \mathbb{Z}, \tag{82}$$

and with which it is fulfilled that $\Phi \in \mathrm{Ord}_2^2(\xi)$.

It is necessary to mention that, for practical purposes, it may be defined that if a fractional iterative method Φ fulfills the properties of the Corollary 1.8 and uses the function (81), it may be called a fractional iterative method accelerated. Finally, it is necessary to mention that fractional iterative methods may be defined in the complex space [24], that is,

$$\{ \Phi(\alpha, x) : \alpha \in \mathbb{R} \backslash \mathbb{Z} \quad \text{and} \quad x \in \mathbb{C}^n \}. \tag{83}$$

However, due to the part of the integral operator that fractional operators usually have, it may be considered that in the matrix $A_{g,\alpha}$ each fractional operator o_k^α is obtained for a real variable $[x]_k$, and if the result allows it, this variable is subsequently substituted by a complex variable $[x_i]k$, that is,

$$A_{g,\alpha}(x_i) := A_{g,\alpha}(x)\big|_{x \longrightarrow xi}, \quad x \in \mathbb{R}^n, \quad x_i \in \mathbb{C}^n. \tag{84}$$

Therefore, it is possible to obtain the following corollaries:

Corollary 1.9 *Let* $f : \Omega \subset \mathbb{C}^n \to \mathbb{C}^n$ *be a function such that* $f \in \mathrm{Ord}^2_{NR}(\xi)$, *let* $g : \mathbb{C}^n \to \mathbb{C}^n$ *be a function such that* $g^{(1)}(x) = f^{(1)}(x)\ \forall x \in B(\xi;\delta)$, *and let* $\Phi : (\mathbb{R}\backslash\mathbb{Z}) \times \mathbb{C}^n \to \mathbb{C}^n$ *be an iteration function given by the equation (62). So, for each operator* $o_x^\alpha \in {}_nO^1_{x,\alpha}(g)$ *such that there exists the matrix* $A^{-1}_{g,\alpha} = A_\alpha(o_x^\alpha) \circ A_\alpha^T(g)$, *it follows that the matrix fulfills the following condition*

$$\lim_{\alpha \to 1} A_{g,\alpha}(x) = \left(f^{(1)}(x)\right)^{-1} \quad \forall x \in B(\xi;\delta). \tag{85}$$

As a consequence, by the Corollary 1.8, if $\Phi(A_{g,\alpha}) \in Ord^1_2(\xi) \Rightarrow \Phi\left(A_{g,\alpha_f}\right) \in Ord^2_2(\xi)$.

Corollary 1.10 *Let* $f : \Omega \subset \mathbb{C}^n \to \mathbb{C}^n$ *be a function such that* $f \in Ord^2_{NR}(\xi)$, *let* $\{g_k\}_{k=1}^N$ *be a finite sequence of functions* $g_k : \mathbb{C}^n \to \mathbb{C}^n$ *such that it defines a finite sequence of operators* $\left\{o^\alpha_{k,x}\right\}_{k=1}^N$ *through the following condition*

$$o^\alpha_{k,x} \in {}_nMO^{\infty,u}_{x,\alpha}(g_k) \quad \forall k \geq 1, \tag{86}$$

and let $\Phi : (\mathbb{R}\backslash\mathbb{Z}) \times \mathbb{C}^n \to \mathbb{C}^n$ be an iteration function given by the Eq. (62). So, if there exists a matrix $A_{N,\alpha}$ such that it fulfills the following conditions

$$\exists A^{-1}_{N,\alpha} = \sum_{k=1}^N A_\alpha\left(o^\alpha_{k,x}\right) \circ A_\alpha^T(g_k) \quad \text{and} \quad \lim_{\alpha \to 1} A_{N,\alpha}(x) = \left(f^{(1)}(x)\right)^{-1} \quad \forall x \in B(\xi;\delta). \tag{87}$$

As a consequence, by the Corollary 1.8, if $\Phi(A_{N,\alpha}) \in \mathrm{Ord}^1_2(\xi) \Rightarrow \Phi\left(A_{N,\alpha_f}\right) \in \mathrm{Ord}^2_2(\xi)$.

6. Conclusions

It is worth mentioning that it is feasible to develop more complex algebraic structures of fractional operators using the presented results. For example, without loss of generality, considering the modified Hadamard product (20) and the operation (36), a commutative and unitary ring of fractional operators may be defined as follows

$${}_m\mathrm{R}(A_\alpha(o_x^\alpha)) := \left({}_m\mathrm{G}(A_\alpha(o_x^\alpha)), \circ, *\right), \tag{88}$$

in which it is not difficult to verify the following properties:

1. The pair $\left({}_m\mathrm{G}(A_\alpha(o_x^\alpha)), \circ\right)$ is an Abelian group.

2. The pair $\left({}_m\mathrm{G}(A_\alpha(o_x^\alpha)), *\right)$ is an Abelian monoid.

3. $\forall A_\alpha^{\circ p}, A_\alpha^{\circ q}, A_\alpha^{\circ r} \in {}_m\mathrm{R}(A_\alpha(o_x^\alpha))$, the operation $*$ is distributive with respect to the operation $\circ$, that is,

$$\begin{cases} A_\alpha^{\circ p} * \left(A_\alpha^{\circ q} \circ A_\alpha^{\circ r}\right) = \left(A_\alpha^{\circ p} * A_\alpha^{\circ q}\right) \circ \left(A_\alpha^{\circ p} * A_\alpha^{\circ r}\right) \\ \left(A_\alpha^{\circ p} \circ A_\alpha^{\circ q}\right) * A_\alpha^{\circ r} = \left(A_\alpha^{\circ p} * A_\alpha^{\circ r}\right) \circ \left(A_\alpha^{\circ q} * A_\alpha^{\circ r}\right) \end{cases}. \tag{89}$$

Author details

A. Torres-Hernandez[1,2*], F. Brambila-Paz[3] and R. Ramirez-Melendez[2]

1 Faculty of Science, Department of Physics, Universidad Nacional Autónoma de México, Mexico City, Mexico

2 Department of Information and Communication Technologies, Music and Machine Learning Lab, Universitat Pompeu Fabra, Barcelona, Spain

3 Faculty of Science, Department of Mathematics, Universidad Nacional Autónoma de México, Mexico City, Mexico

*Address all correspondence to: anthony.torres@ciencias.unam.mx

References

[1] De Oliveira EC, Machado JAT. A review of definitions for fractional derivatives and integral. Mathematical Problems in Engineering. 2014;**2014**:1-6

[2] Teodoro GS, Machado JAT, De Oliveira EC. A review of definitions of fractional derivatives and other operators. Journal of Computational Physics. 2019;**388**:195-208

[3] Valério D, Ortigueira MD, Lopes AM. How many fractional derivatives are there? Mathematics. 2022;**10**:737

[4] Hilfer R. Applications of Fractional Calculus in Physics. Singapore: World Scientific Publishing Co. Pte. Ltd.; 2000. pp. 3-73

[5] Oldham K, Spanier J. The Fractional Calculus Theory and Applications of Differentiation and Integration to Arbitrary Order. London: Academic Press, Inc.; 1974. pp. 25-121

[6] Kilbas AA, Srivastava HM, Trujillo JJ. Theory and Applications of Fractional Differential Equations. Amsterdam: Elsevier; 2006. pp. 69-132

[7] Torres-Hernandez A, Brambila-Paz F, Iturrarán-Viveros U, Caballero-Cruz R. Fractional newton-raphson method accelerated with aitken's method. Axioms. 2021;**10**(2):1-25. DOI: 10.3390/axioms10020047

[8] Thomas J. Leibniz rule for fractional derivatives generalized and an application to infinite series. SIAM Journal on Applied Mathematics. 1970; **18**(3):658-674

[9] Almeida R. A caputo fractional derivative of a function with respect to another function. Communications in Nonlinear Science and Numerical Simulation. 2017;**44**:460-481

[10] Hui F, Wu G-C, Yang G, Huang L-L. Continuous time random walk to a general fractional fokker–planck equation on fractal media. The European Physical Journal Special Topics. 2021;**2021**:1-7

[11] Fan Q, Wu G-C, Hui F. A note on function space and boundedness of the general fractional integral in continuous time random walk. Journal of Nonlinear Mathematical Physics. 2022;**29**(1):95-102

[12] Abu-Shady M, Kaabar MKA. A generalized definition of the fractional derivative with applications. Mathematical Problems in Engineering. 2021;**2021**:1-9

[13] Khaled M. New fractional derivative with non-singular kernel for deriving legendre spectral collocation method. Alexandria Engineering Journal. 2020; **2020**:1909-1917

[14] Rahmat MRS. A new definition of conformable fractional derivative on arbitrary time scales. Advances in Difference Equations. 2019;**2019**(1):1-16

[15] Vanterler J et al. On the hilfer fractional derivative. Communications in Nonlinear Science and Numerical Simulation. 2018;**60**:72-91

[16] Jarad F, Uğurlu E, Abdeljawad T, Baleanu D. On a new class of fractional operators. Advances in Difference Equations. 2017;**2017**(1):1-16

[17] Atangana A, Gómez-Aguilar JF. A new derivative with normal distribution kernel: Theory, methods and applications. Physica A: Statistical Mechanics and Its Applications. 2017;**476**:1-14

[18] Yavuz M, Özdemir N. Comparing the new fractional derivative operators involving exponential and mittag-leffler kernel. Discrete & Continuous Dynamical Systems-S. 2020;**13**:995

[19] Liu J-G, Yang X-J, Feng Y-Y, Cui P. New fractional derivative with sigmoid function as the kernel and its models. Chinese Journal of Physics. 2020;**68**: 533-541

[20] Yang X-J, Machado JAT. A new fractional operator of variable order: Application in the description of anomalous diffusion. Physica A: Statistical Mechanics and its Applications. 2017;**481**:276-283

[21] Atangana A. On the new fractional derivative and application to nonlinear fisher's reaction–diffusion equation. Applied Mathematics and Computation. 2016;**273**:948-956

[22] He J-H, Li Z-B, Wang Q-l. A new fractional derivative and its application to explanation of polar bear hairs. Journal of King Saud University-Science. 2016;**28**(2):190-192

[23] Sene N. Fractional diffusion equation with new fractional operator. Alexandria Engineering Journal. 2020; **59**(5):2921-2926

[24] Torres-Hernandez A, Brambila-Paz F. Sets of fractional operators and numerical estimation of the order of convergence of a family of fractional fixed-point methods. Fractal and Fractional. 2021;**5**(4):240

[25] Torres-Hernandez A, Brambila-Paz F, Montufar-Chaveznava R. Acceleration of the order of convergence of a family of fractional fixed point methods and its implementation in the solution of a nonlinear algebraic system related to hybrid solar receivers. Applied Mathematics and Computation. 2022;**429**:127231. DOI: 10.1016/j.amc.2022.127231

[26] Torres-Hernandez A. Code of a multidimensional fractional quasi-newton method with an order of convergence at least quadratic using recursive programming. Applied Mathematics and Sciences: An International Journal (MathSJ). 2022;**9**: 17-24. DOI: 10.5121/mathsj.2022.9103

[27] Safdari-Vaighani A, Heryudono A, Larsson E. A radial basis function partition of unity collocation method for convection–diffusion equations arising in financial applications. Journal of Scientific Computing. 2015;**64**(2): 341-367

[28] Torres-Hernandez A, Brambila-Paz F, Torres-Martínez C. Numerical solution using radial basis functions for multidimensional fractional partial differential equations of type black–scholes. Computational and Applied Mathematics. 2021;**40**(2):248

[29] Traore A, Sene N. Model of economic growth in the context of fractional derivative. Alexandria Engineering Journal. 2020;**59**(6): 4843-4850

[30] Guariglia E. Fractional calculus, zeta functions and shannon entropy. Open Mathematics. 2021;**19**(1):87-100

[31] Torres-Henandez A, Brambila-Paz F. An approximation to zeros of the riemann zeta function using fractional calculus. Mathematics and Statistics. 2021;**9**(3):309-318

[32] Vega E D-l, Torres-Hernandez A, Rodrigo PM, Brambila-Paz F. Fractional derivative-based performance analysis of hybrid

thermoelectric generator-concentrator photovoltaic system. Applied Thermal Engineering. 2021;**193**:1169

[33] Torres-Hernandez A, Brambila-Paz F, Rodrigo PM, De-la-Vega E. Reduction of a nonlinear system and its numerical solution using a fractional iterative method. Journal of Mathematics and Statistical Science. 2020;**6**:285-299

[34] Erfanifar R, Sayevand K, Esmaeili H. On modified two-step iterative method in the fractional sense: Some applications in real world phenomena. International Journal of Computer Mathematics. 2020; **97**(10):2109-2141

[35] Cordero A, Girona I, Torregrosa JR. A variant of chebyshev's method with 3th-order of convergence by using fractional derivatives. Symmetry. 2019; **11**(8):1017

[36] Torres-Hernandez A, Brambila-Paz F. Fractional newton-raphson method. Applied Mathematics and Sciences: An International Journal (MathSJ). 2021;**8**:1-13. DOI: 10.5121/mathsj.2021.8101

[37] Gdawiec K, Kotarski W, Lisowska A. Newton's method with fractional derivatives and various iteration processes via visual analysis. Numerical Algorithms. 2021;**86**(3):953-1010

[38] Gdawiec K, Kotarski W, Lisowska A. Visual analysis of the newton's method with fractional order derivatives. Symmetry. 2019;**11**(9):1143

[39] Akgül A, Cordero A, Torregrosa JR. A fractional newton method with 2th-order of convergence and its stability. Applied Mathematics Letters. 2019;**98**: 344-351

[40] Candelario G, Cordero A, Torregrosa JR. Multipoint fractional iterative methods with (2+ 1) th-order of convergence for solving nonlinear problems. Mathematics. 2020;**8**:452

[41] Candelario G, Cordero A, Torregrosa JR, Vassileva MP. An optimal and low computational cost fractional newton-type method for solving nonlinear equations. Applied Mathematics Letters. 2022;**124**:107650

Chapter 4

Total and Partial Differentials as Algebraically Manipulable Entities

Maria Isabelle Fite and Jonathan Bartlett

Abstract

Differential operators usually result in derivatives expressed as a ratio of differentials. For all but the simplest derivatives, these ratios are typically not algebraically manipulable, but must be held together as a unit in order to prevent contradictions. However, this is primarily a notational and conceptual problem. The work of Abraham Robinson has shown that there is nothing contradictory about the concept of an infinitesimal differential operating in isolation. In order to make this system extend to all of calculus, however, some tweaks to standard calculus notation are required. Understanding differentials in this way actually provides a more straightforward understanding of all of calculus for students, and minimizes the number of specialized theorems students need to remember, since all terms can be freely manipulated algebraically.

Keywords: differentials, differential operators, derivatives, partial derivatives, total derivatives

1. Introduction

Derivatives are usually written in a notation, such as $\frac{dy}{dx}$, where the notation implies that there are two distinct values, dy and dx, at play. Historically, dy and dx were considered infinitesimal values—values so small that they are practically zero, but not quite zero, and often became real numbers when put in ratio with each other. This understanding was challenged by practitioners who thought that infinitesimal values were insufficiently rigorous to be used in mathematics.

This led to a reconsideration of derivatives using the concept of a limit. In the limit definition of the derivative, the dy and dx terms do not have independent existences, but exist only within the ratio itself. In this conception, the ratio is merely suggestive of how the derivative was originally produced but does not represent an actual quotient of two distinct values. The limit definition of the derivative has been reinforced by the fact that treating differentials as distinct values leads to contradictions in many cases.

However, the work of Abraham Robinson in the 1960s showed that there was no fundamental flaw in expanding the number system to include infinitesimals. The hyperreal numbers are an extension of the real numbers which allows for infinitesimals and infinities to be constructed in a manner equally rigorous with the real numbers. Additionally, unlike other conceptions of infinities, the hyperreal numbers have an additional advantage that infinitesimals and infinities can be manipulated using arithmetic and algebraic operations.

However, if infinitesimals can be readily considered without contradiction, why does the notation for derivative operations often lead to contradiction? The flaw here is actually in the notation itself. Because the notation was not considered factual but merely suggestive, practitioners tended to ignore the problematic cases rather than solve them. By considering new and more rigorous approaches to notation, a better notation can be developed which includes infinitesimal values, removes the contradictions, and provides a more straightforward understanding of differential notation and formulas. In these new formulations, differentials such as $\mathrm{d}y$ and $\mathrm{d}x$ are fully independent, algebraically manipulable entities.

2. Problem of separating differentials in modern Leibniz notation

While the problems that occur when trying to separate differentials in modern Leibniz notation are well-known, it is worth revisiting them briefly. First of all, it is interesting to note that there are essentially no inconsistencies or contradictions when dealing with first-order total differentials. For instance, taking the equation $y = x^3$, the derivative is $\frac{\mathrm{d}y}{\mathrm{d}x} = 3x^2$. Since the derivative of the inverse function is $\frac{\mathrm{d}x}{\mathrm{d}y}$, this can be found simply by inverting both sides of the equation, so that $\frac{\mathrm{d}x}{\mathrm{d}y} = \frac{1}{\frac{\mathrm{d}y}{\mathrm{d}x}} = \frac{1}{3x^2}$. Likewise, integrating is often preceded by multiplying both sides by a differential, so that $\frac{\mathrm{d}y}{\mathrm{d}x} = 3x^2$ becomes $\mathrm{d}y = 3x^2\,\mathrm{d}x$.

The problems become more apparent on higher-order derivatives. The typical notation for the second derivative of $y = x^3$ is $\frac{\mathrm{d}^2y}{\mathrm{d}x^2} = 6x$. However, if the notation were taken seriously, this would be seen as a quotient of the higher-order differential d^2y and the square of $\mathrm{d}x$. Doing this, however, would break the chain rule. For instance, if you had $x = t^2$, then you could calculate $\frac{\mathrm{d}^2y}{\mathrm{d}t^2}$ by simply multiplying $\frac{\mathrm{d}^2y}{\mathrm{d}x^2}$ by $\left(\frac{\mathrm{d}x}{\mathrm{d}t}\right)^2$. Doing so, however, yields an incorrect second derivative of $\frac{\mathrm{d}^2y}{\mathrm{d}t^2} = 24t^4$ rather than the correct $\frac{\mathrm{d}^2y}{\mathrm{d}t^2} = 30t^4$. This is normally calculated using the chain rule for the second derivative (or higher derivatives using Fa'a di Bruno's formula [1]). While the second derivative chain rule works, it provides no algebraic intuition for why it works, and seems to be in conflict with the idea of treating differentials as separable values.

Dealing with partial derivatives brings up innumerable problematic cases even for the first derivative. If f is a function of x and y, and x and y are both functions of t, then the total derivative of f with respect to t is $\frac{\mathrm{d}f}{\mathrm{d}t} = \frac{\partial f}{\partial x}\frac{\mathrm{d}x}{\mathrm{d}t} + \frac{\partial f}{\partial y}\frac{\mathrm{d}y}{\mathrm{d}t}$. Since x is a function of one variable, $\frac{\partial x}{\partial t} = \frac{dx}{dt}$ (likewise for y). Then the equation becomes $\frac{\mathrm{d}f}{\mathrm{d}t} = \frac{\partial f}{\partial x}\frac{\partial x}{\partial t} + \frac{\partial f}{\partial y}\frac{\partial y}{\partial t}$. Treating the partial differentials as distinct values, this reduces to $\frac{\mathrm{d}f}{\mathrm{d}t} = \frac{\partial f}{\partial t} + \frac{\partial f}{\partial t}$. [1] Now

[1] A possible objection is that the ∂x in $\frac{\partial f}{\partial x}$ may not be the same infinitesimal as the ∂x in $\frac{\partial x}{\partial t}$. However, the value of ∂f depends on the value of the ∂x in $\frac{\partial f}{\partial x}$, and the value of the ∂x in $\frac{\partial x}{\partial t}$ depends on ∂t. So one could choose the ∂xs to be equal, and the values of ∂f and ∂t would adjust accordingly, leaving the values of $\frac{\partial f}{\partial x}$ and $\frac{\partial x}{\partial t}$ unchanged.

that it is expressed in terms of a single variable, $\frac{df}{dt} = \frac{\partial f}{\partial t}$, so this yields $\frac{df}{dt} = \frac{df}{dt} + \frac{df}{dt} = 2\frac{df}{dt}$. Dividing both sides by $\frac{df}{dt}$ yields the contradiction $1 = 2$.

As will be described, the issues in these problematic cases stem from deficiencies in the notation, not deficiencies in the concept of differentials as infinitesimals nor in the idea that differentials can be considered independently of each other. By taking a more rigorous approach to the development of the notation of higher order derivatives and partial derivatives, a straightforward notation can be obtained which enables differentials to be considered as fully distinct values.

3. Historical formal definitions of the derivative

The derivative of a function measures how the function changes as the independent variable varies. For instance, if the derivative of a function $f(x)$ is 3 when $x = 5$, that means $f(x)$ is increasing at a rate of 3 units up to every 1 unit across whenever x is 5. Another way to say the same information is that the function's slope at $x = 5$ is $3/1 = 3$.

Normally, slope is defined with reference to two points. When measuring velocity, for instance, which is the ratio of the change in position to the change in time, one would measure two different times with their positions and compare them. The derivative attempts to calculate the slope using only one point together with an equation. Since only one point is used, the change in x is infinitely small, and so is the change in y. Different ways of dealing with these infinities lead to different formal definitions of the derivative.

3.1 Newton's definition

Isaac Newton provided one of the first definitions of a derivative in his book *Methodus fluxionum et serierum infinitarum*, or "The Method of Fluxions and Infinite Series" in English [2, 3]. Newton thought of his graphs as being drawn over time, with the x-coordinate increasing at a constant speed while the rate of increase in the y-coordinate varied. A variable's rate of change with respect to time (what we would now call a derivative with respect to time) was called a "fluxion," which was denoted by applying a dot above a variable, such as $\dot{x}$ (which represents the derivative of x with respect to time) [3].

To avoid having to define an infinitely small quantity, Newton worked with full derivatives, ratios of infinitesimals. Since Newton assumed all his variables depended on time, he could then switch out the infinitesimal change in x and change in y for the change in x over time and the change in y over time, which were both real numbers. The ratio remained the same, and the infinities were avoided [3].

3.2 Leibniz's definition

Unlike Newton, Gottfried Leibniz preferred to consider the change in x and the change in y separately. He used the notation dx for an infinitesimal difference in x and dy/dx for a ratio of infinitesimals, which represented the slope of a curve at a point. Leibniz considered d an operator, with $dx = d(x)$ being the output of d acting on the variable x. This allowed him to apply d more than once, resulting in $d^2x = d(d(x))$,

$d^3x = d(d(d(x)))$, and so on. Just like dx was infinitely smaller than x, Leibniz said d^nx was infinitely smaller than $d^{n-1}x$ [3].

Although his calculus relied on the concept of an infinitesimal, Leibniz regarded infinitesimals as only "purely ideal entities... useful fictions, introduced to shorten arguments and aid insight" [3]. However, Leibniz was never able to rigorously define his infinitesimals nor how they behaved. Therefore, while they seemed to work well, the lack of clarity caused some skeptics to regarded them with suspicion, ridiculing them as "ghosts of departed quantities" [4].

3.3 Delta-epsilon (limit) definition

Concerns about the fishy nature of infinitesimals, treated like nonzero numbers when dividing but also like zero when adding, led to the reformulation of calculus using the idea of limits. The limit of $f(x)$ as x approaches a is the value $f(x)$ approaches as x becomes closer to a.

More precisely, the limit of $f(x)$ as x approaches a is L if for any given positive number ε there is a corresponding positive number δ such that the difference between $f(x)$ and L is less than ε whenever the difference between x and a is less than δ [5].

Limits can then be used to define the derivative of a function $f(x)$ as

$$f'(x) = \lim_{h \to 0} \frac{f(x+h) - f(x)}{h} \tag{1}$$

When limits are used to define a derivative, it makes no sense to pull apart the change in x and the change in y, as both the limit of the numerator and the limit of the denominator evaluate to zero, and division by zero is undefined.

4. Hyperreal numbers and the definition of the derivative

While the limit definition of a derivative solves the philosophical problems of infinitesimals, it does not allow the change in y to be separated from the change in x. This led Abraham Robinson to return to Leibniz's infinitesimals in 1958, putting them on a new set-theoretic foundation and creating the field of nonstandard analysis [3].

While there are different ways to construct hyperreal numbers, the approach we will take here is based on the set theory approach described by Herrmann in [6], with many of the definitions taken from there as well. We will begin by describing hyperreal numbers (including infinitesimals), and then describe the differential operator as being an operator that can be applied using infinitesimals.

For defining the infinitesimals, the core idea is to take the set of all infinitely long sequences of real numbers, denoted $\mathbb{R}^{\mathbb{N}}$. Some of these sequences match other sequences so closely they can be considered equivalent. Each real number is then assigned to a set of equivalent sequences. Then, some of the remaining sets of equivalent sequences can be assigned to infinitesimals. Finally, all the operations normally done on real numbers can be translated to operations between sets of equivalent sequences.

4.1 Filters, the cofinite filter, and free ultrafilters: Defining big enough

A *filter* provides a way to classify subsets of a set as either big enough or not big enough.

Let X be a nonempty set. A nonempty subset F of the set of all subsets of X is a proper filter on X if and only if:

$$\text{(i) for each } A, B \in F, A \cap B \in F \tag{2}$$

$$\text{(ii) if } A \subset B \subset X \text{ and } A \in F, \text{then } B \in F \tag{3}$$

$$\text{(iii) } \varnothing \notin F \tag{4}$$

The *cofinite filter* C is defined as

$$C = x \mid (x \subset X) \text{ and } (X - x) \text{ is finite} \tag{5}$$

where X is an infinite set. C is called the cofinite filter because a subset x of X gets to be in the filter C if and only if X without x is a finite set. C gives a mathematical way to define whether an infinite set is considered big enough.

For instance, if C is the cofinite filter on $\mathbb{R}$, the real numbers, the set of all integers $\mathbb{Z}$ is not big enough to be in C, even though it is an infinite subset of $\mathbb{R}$, because there are infinitely many real numbers that are not integers. However, $\mathbb{R}^*$, the real numbers excluding zero, is big enough to be a member of C, because there is only one real number, zero, that is not in the real numbers excluding zero.

An *ultrafilter* is the biggest filter on a given infinite set X. An ultrafilter that has C as a subset is called a *free ultrafilter*.

4.2 Equivalence classes of $\mathbb{R}^{\mathbb{N}}$: Classifying equivalent sequences together

Let $\mathbb{R}^{\mathbb{N}}$ represent the set of all sequences with domain $\mathbb{N}$ and range values in $\mathbb{R}$. Let A and B be two sequences in $\mathbb{R}^{\mathbb{N}}$. A is said to be equivalent to B ($A =_U B$) if a sufficiently large number of their elements match, or

$$A =_U B \Leftrightarrow n \mid \{A_n = B_n\} = S \in U \tag{6}$$

The free ultrafilter U determines whether the set of matching elements is big enough.

This relation $=_U$ is an equivalence relation on $\mathbb{R}^{\mathbb{N}}$, so it can partition $\mathbb{R}^{\mathbb{N}}$ into equivalence classes. Each equivalence class $[A]$ contains all the sequences in $\mathbb{R}^{\mathbb{N}}$ that are equivalent to A, including A itself.

The set of all these equivalence classes is called the set of the hyperreal numbers, denoted $^*\mathbb{R}$.

4.3 Connecting the real numbers to the hyperreals

We can define a function f that takes each $x \in \mathbb{R}$ and gives the unique $[R]$, where $\{n \mid R_n = x\} \in U$. This function f assigns to each real number x a hyperreal number $[R]$, namely that set of all sequences where a sufficiently large number of each sequence's elements is x. Often, $f(x)$ is represented by $^*x$. For instance, the hyperreal ^{*3} is the set of all sequences equivalent ($=_U$) to $\{3, 3, 3, \ldots\}$.

Most applications of math use real numbers, so it is helpful to define the subset of the hyperreals that corresponds to the real numbers. The image of a subset X of $\mathbb{R}$ under f is denoted $^\sigma X$. Each hyperreal number $^*x$ in $^\sigma X$ corresponds to a real number x in X. Since $\mathbb{R}$ is a subset of $\mathbb{R}$, $^\sigma\mathbb{R}$ is the subset of the hyperreals that corresponds to the real numbers.

4.4 Operations on the hyperreals

In order for algebra in $^*\mathbb{R}$ to replace algebra in the real numbers, operations like + and $\cdot$, among others, have to be defined between members of $^*\mathbb{R}$. It is also useful to define the relation $\leq$ and the absolute value function.

Let a, b, and c be elements of $^*\mathbb{R}$, and let $^*+ : {}^*\mathbb{R} \to {}^*\mathbb{R}$ be defined as

$$a^* + b = c \Leftrightarrow \{n \mid A_n + B_n = C_n\} \in U \tag{7}$$

for any $A_n \in a$, $B_n \in b$, and $C_n \in c$. That is, the sum of 2 elements of $^*\mathbb{R}$, a and b, are equal to another element of $^*\mathbb{R}$, c, if and only if a sufficiently large number of the elements of the sequences $A_n + B_n$ and C_n match, for any sequence A_n in a, B_n in b, and C_n in c. Hyperreal multiplication ($^*\cdot$) can be defined similarly.

To construct a hyperreal greater than relation, for each $a = [A], b = [B] \in {}^*\mathbb{R}$ define

$$a^* \leq b \Leftrightarrow \{n \mid A_n \leq B_n\} \in U \tag{8}$$

$a^* \leq b$ if and only if, given any sequence in a and any sequence in b, a sufficiently large number of elements in a's sequence are less than or equal to their corresponding elements in b's sequence.

These operations establish the structure $\langle {}^*\mathbb{R}, {}^*+, {}^*\cdot, {}^*\leq \rangle$ as a totally ordered field, with $[0]$ as the identity for $^*+$ and $[1]$ as the identity for $^*\cdot$ ([6], p. 11).

Finally, the absolute value function can be defined for members of $a \in {}^*\mathbb{R}$ with

$$^*|a| = |a| = b \Leftrightarrow \{n \mid A_n| = B_n|\} \in U \tag{9}$$

The absolute value of a hyperreal number a is a hyperreal number b if and only if, given a sequence in a and a sequence in b, a sufficiently large number of elements in b's sequence match the absolute value of their corresponding elements in a's sequence.

In summary, $+$, $\cdot$, $\leq$ and the absolute value function, which are defined on the real numbers, can be translated to operations on the hyperreal numbers.

4.5 Infinitesimals in the hyperreals

Not all of the members of $^*\mathbb{R}$ correspond to real numbers, because not all sequences of real numbers are constant sequences. Some of the remaining hyperreals correspond to infinitesimals.

A hyperreal number a is infinitely large if

$$^*x < |a| \text{for each} ^*x \in {}^\sigma\mathbb{R} \tag{10}$$

or in other words, if its absolute value is bigger than every hyperreal that corresponds to a real number.

A hyperreal number b is an infinitesimal or as Newton stated infinitely small if

$$0 \leq |b| < {}^*x \text{ for each } 0 < x \in \mathbb{R}. \tag{11}$$

Similarly, a hyperreal is an infinitesimal if its absolute value is bigger than or equal to ^{*0} and yet smaller than every hyperreal that corresponds to a positive real number.

Notice that ^{*0}, which is the equivalence class that contains $\{0,0,0, \ldots\}$, is the trivial infinitesimal.

For a nontrivial example of an infinitesimal, consider the equivalence class g containing the sequence $\{0, 1, \frac{1}{2}, \frac{1}{3}, \frac{1}{4} \dots\}$. "Then $g \neq^* 0$. Now for each $x \in \mathbb{R}^+$ there is some $m \in \mathbb{N}$, $m \neq 0$ such that $0 < \frac{1}{m} < x$. Thus $^*0 < \frac{^*1}{^*m} < {}^*x$. ... [and] g is an infinitesimal" ([6], p. 17).

4.6 Division with infinitesimals

If infinitesimals are smaller than every real number, can you still divide by them?

Consider a nonzero infinitesimal, say ε, and a sequence in ε, say A. Even if some of A's elements are zeros, $\varepsilon \neq^* 0$, so the set of all zeros in A is not big enough to be in the ultrafilter U. So, the nonzero elements of A *are* in U, since U is an ultrafilter. It is then possible to define another sequence B where $B_n = \frac{1}{A_n}$ if $A_n \neq 0$ and $B_n = 0$ if $A_n = 0$. B satisfies the property $[A]^* \cdot [B] = [1]$, and so $[B]$ is the multiplicative inverse of $[A]$.

In summary, even if there are sequences in ε with zeros, $\frac{[1]}{\varepsilon}$ is still defined, and so it is still possible to divide by ε ([6], p. 11).

4.7 The standard and principal part functions

Hyperreal expressions can be converted into real expressions using the standard part function, st(), which yields the closest real number to the hyperreal expression. The standard part of an infinitesimal number is always zero. For infinite values, the standard part yields $+\infty$ or $-\infty$, which is the non-specific infinity indicating that the value is out of range of the real numbers.

The principal part function, pt(), will yield the most significant component of a hyperreal expression [7]. In a hyperreal expression, imagine ω representing a benchmark infinite value, with $\varepsilon = \frac{1}{\omega}$ representing an associated benchmark infinitesimal. The hyperreal expression $-2\omega^2 + \omega - 5 + 3\varepsilon$ represents four different orders of infinity. The most significant one is $-2\omega^2$, and, thus, it is the principal part. For the infinitesimal expression $5\varepsilon^2 + \varepsilon^3$, $5\varepsilon^2$ is the principal part.

The principal part of a hyperreal expression is important because non-principal parts, being infinitely less significant than the principal part by definition, do not affect the large-scale behaviors of smooth and continuous functions.

4.8 Differentials and derivatives using hyperreals

The derivative of a function $y = f(x)$ using the hyperreals is denoted $\frac{\mathrm{d}y}{\mathrm{d}x}$, the change in y divided by the change in x, just like using Leibniz's notation. However, we can actually define the differentials themselves as infinitesimals, without referring to ratios.

Many have a hard time conceiving of just what a differential is and means. It is easy enough to say that a differential is an infinitesimal, but how exactly are individual differentials defined, especially when not being examined in the context of a derivative? What exactly does the higher-order notation d^2y mean?

Let us first remember that, in order to be in a relation, two (or more) variables have to be related to each other in some way. Therefore, we can imagine some variable, let us call it q, not explicitly mentioned in the equation, which is in some sense the "ultimate" independent variable.

Note that this variable does not need to be explicitly defined. In fact, it is better if it is not defined explicitly. The reason for this is that defining q explicitly means that there is some chance that there exists yet another deeper, more fundamental variable. What we are looking for is the deepest, most fundamental, most independent variable. Keeping q as a hypothetical independent variable means that our reasoning will continue to hold in the face of finding more and more fundamental quantities. Our reasoning about an *actual* variable may fail to hold if it is found to not be the fundamental quantity. We will imagine q to be smoothly increasing by the infinitesimal ε.

Since q is the ultimate variable that relates every other variable in the equation, every variable can (theoretically) be written in terms of q. y is actually shorthand for $y(q)$, x is a shorthand for $x(q)$, and so on. We can then define the differential of an expression (including just a variable) to be the simple difference between the expression at some value $q + \varepsilon$ and the expression at some value q. When taking the differential of a variable, we will use the shorthand $\mathrm{d}y$ to mean $\mathrm{d}(y)$.

$$\mathrm{d}y = \mathrm{d}(y) = y(q + \varepsilon) - y(q) \tag{12}$$

Note that $\mathrm{d}y$ is also a function of q (this fact will become useful when finding the second differential). Additionally, assuming that y is a smooth and continuous function of q, an infinitesimal change in q will lead to an infinitesimal change in in y, so $\mathrm{d}y$ will also be infinitesimal.

We can also rearrange (12) and obtain

$$y(q + \varepsilon) = y(q) + \mathrm{d}y \tag{13}$$

These definitions provide a generic definition for the differential and consequent manipulation techniques that can be applied to any expression. Let us take the simple example $y = x^2$ (which is $y(q) = x(q)^2$) and apply this differential operator to it. We will also apply the principal part function at the end in order to simplify the expression to its most consequential portion.

$$\begin{aligned}
y &= x^2 & & \\
\mathrm{d}(y) &= \mathrm{d}(x^2) & & \text{differential operator} \\
y(q+\varepsilon) - y(q) &= x(q+\varepsilon)^2 - x(q)^2 & & \text{applying (12)} \\
\mathrm{d}y &= (x(q) + \mathrm{d}x)^2 - x(q)^2 & & \text{applying (13)} \\
\mathrm{d}y &= x(q)^2 + 2x(q)\,\mathrm{d}x + \mathrm{d}x^2 - x(q)^2 & & \text{simplifying} \\
\mathrm{d}y &= 2x(q)\,\mathrm{d}x + \mathrm{d}x^2 & & \\
\mathrm{d}y &= 2x(q)\,\mathrm{d}x & & \text{principal part} \\
\mathrm{d}y &= 2x\,\mathrm{d}x & & \text{shorthand}
\end{aligned}$$

The second differential is the same process. It is merely the differential operator applied where differentials are concerned. $\mathrm{d}y$ is actually $\mathrm{d}(y(q)))$, but we will refer to it as $\mathrm{d}y(q)$ and $\mathrm{d}y(q + \varepsilon)$ for a compromise of brevity and clarity. The notation d^2y will likewise be shorthand for $\mathrm{d}(\mathrm{d}(y(q)))$.

$$
\begin{aligned}
\mathrm{d}y &= 2x\,\mathrm{d}x \\
\mathrm{d}(\mathrm{d}y) &= \mathrm{d}(2x\,\mathrm{d}x) && \text{differential operator} \\
&= 2x(q+\varepsilon)\,\mathrm{d}x(q+\varepsilon) - 2x(q)\,\mathrm{d}x(q) && \text{applying(12)} \\
&= 2(x(q)+\mathrm{d}x(q))(\mathrm{d}x(q)+\mathrm{d}(\mathrm{d}x(q))) - 2x(q)\,\mathrm{d}x(q) && \text{applying (13)} \\
&= 2x(q)\,\mathrm{d}x(q) + 2x(q)\mathrm{d}(\mathrm{d}x(q)) && \text{simplifying} \\
&\quad +2\,\mathrm{d}x(q)^2 + 2\mathrm{d}x\,\mathrm{d}(\mathrm{d}x(q)) - 2x(q)\,\mathrm{d}x(q) \\
&= 2x(q)\mathrm{d}(\mathrm{d}x(q)) + 2\mathrm{d}x(q)^2 + 2\mathrm{d}x\,\mathrm{d}(\mathrm{d}x(q)) \\
&= 2x(q)\mathrm{d}(\mathrm{d}x(q)) + 2\,\mathrm{d}x(q)^2 && \text{principal part} \\
\mathrm{d}^2y &= 2x\,\mathrm{d}^2x + 2\,\mathrm{d}x^2 && \text{shorthand}
\end{aligned}
$$

This second differential will typically be a second order infinitesimal. The process can be further repeated for higher order differentials.

The $2x\,\mathrm{d}^2x$ term here may be surprising, but the reason for it will become clear in Section 5 when we eliminate the contradictions present in the standard notation for higher-order differentials.

Since all variables in the equation are related to each other, they also share some relationship to q. Therefore, the definition of a differential can be defined universally within an equation without taking into account the specifics of the variables encountered.

Ultimately, taking the differential of a function results in a $\mathrm{d}y$, $\mathrm{d}x$, or some other term. However, these terms' definitions are ultimately rooted in this ultimate independent variable q, and the results of incrementing it by some hyperreal infinitesimal ε.

The derivative, then, is simply a ratio of differentials defined in this way. While the terminology of "taking the derivative with respect to x" can still be used, there is no longer anything special about taking the derivative with respect to a variable as opposed to simply dividing by that variable's differential. Additionally, this expands the ability to take total differentials straightforwardly into multivariable situations, providing that all variables can be, in principle, tied back to some underlying construct like q.

5. Extending the total derivative's algebraic manipulability

The hyperreal definition of the derivative has several advantages. Once hyperreal numbers are defined, the definition of the derivative arises naturally from considering the change in a function when its (theoretical) independent variable changes infinitesimally. Unlike the limit definition, the change in y and the change x are separate entities. Using hyperreal numbers, we can rigorously define these entities so that they are manipulable using standard algebraic operators.

However, this requires that we rethink some of the notations from first principles. First of all, now that $\mathrm{d}y$ and $\mathrm{d}x$ are reified entities, they now must be considered in applying such rules as the product rule and the quotient rule. This is straightforward, and the rules are identical to normal calculus rules. The differential of $x^2\,\mathrm{d}x$ is the result of applying the product rule to the product of x^2 and $\mathrm{d}x$, namely $2x\,\mathrm{d}x^2 + x^2\,\mathrm{d}^2x$.

When this is taken into account, differentials of any order become algebraically manipulable.

5.1 The second derivative

Before taking this idea of algebraically manipulable differentials too far, we need to note that the standard notation for the second derivative, $\frac{d^2y}{dx^2}$, does not work in this manner. The problem, here, is that it implies an improper order of operations [8].

Order of operations is very important when doing derivatives. When doing a derivative, one *first* takes the differential and *then* divides by dx. The second derivative is the derivative of the first, so the next differential occurs *after the first derivative is complete*, and the process finishes by dividing by dx again.

However, what does it look like to take the differential of the first derivative? Basic calculus rules tell us that the quotient rule should be used:

$$\begin{aligned} d\left(\frac{dy}{dx}\right) &= \frac{dx(d(dy)) - dy(d(dx))}{(dx)^2} \\ &= \frac{d^2y}{dx} - \frac{dy}{dx}\frac{d^2x}{dx} \end{aligned}$$

Then, for the second step, this can be divided by dx, yielding:

$$\frac{d\left(\frac{dy}{dx}\right)}{dx} = \frac{d^2y}{dx^2} - \frac{dy}{dx}\frac{d^2x}{dx^2} \tag{14}$$

This, in fact, yields a notation for the second derivative which is equally algebraically manipulable as the first derivative. It is not very pretty or compact, but it works algebraically.

The chain rule for the second derivative fits this algebraic notation correctly, provided we replace each instance of the second derivative with its full form (cf. (30)):

$$\frac{d^2y}{dt^2} - \frac{dy}{dt}\frac{d^2t}{dt^2} = \left(\frac{d^2y}{dx^2} - \frac{dy}{dx}\frac{d^2x}{dx^2}\right)\left(\frac{dx}{dt}\right)^2 + \frac{dy}{dx}\left(\frac{d^2x}{dt^2} - \frac{dx}{dt}\frac{d^2t}{dt^2}\right) \tag{15}$$

This in fact works out perfectly algebraically.[2]

5.2 Higher order derivatives

The notation for the third and higher derivatives can be found using the same techniques as for the second derivative. To find the third derivative of y with

---

[2] Some may be concerned that, in the formula presented in (14), the ratio $\frac{d^2x}{dx^2}$ reduces to zero. However, this is not necessarily true. The concern is that, since $\frac{dx}{dx}$ is always 1 (i.e., a constant), then $\frac{d^2x}{dx^2}$ should be zero. The problem with this concern is that we are no longer taking $\frac{d^2x}{dx^2}$ to be the derivative of $\frac{dx}{dx}$. Using the notation in (14), the derivative of $\frac{dx}{dx}$ would be:

$$\frac{d\left(\frac{dx}{dx}\right)}{dx} = \frac{d^2x}{dx^2} - \frac{dx}{dx}\frac{d^2x}{dx^2} \tag{16}$$

In this case, since $\frac{dx}{dx}$ reduces to 1, the expression is self-evidently zero. However, in (16), the term $\frac{d^2x}{dx^2}$ is not itself necessarily zero, since it is *not* the second derivative of x with respect to x.

respect to x, one starts with the second derivative, takes the differential, and divides by $\mathrm{d}x$:

$$\frac{\mathrm{d}\left(\frac{\mathrm{d}\left(\frac{\mathrm{d}y}{\mathrm{d}x}\right)}{\mathrm{d}x}\right)}{\mathrm{d}x} = \frac{\mathrm{d}\left(\frac{\mathrm{d}^2y}{\mathrm{d}x^2} - \frac{\mathrm{d}y}{\mathrm{d}x}\frac{\mathrm{d}^2x}{\mathrm{d}x^2}\right)}{\mathrm{d}x} = \frac{\mathrm{d}^3y}{\mathrm{d}x^3} - \frac{\mathrm{d}y}{\mathrm{d}x}\frac{\mathrm{d}^3x}{\mathrm{d}x^3} - 3\frac{\mathrm{d}^2x}{\mathrm{d}x^2}\frac{\mathrm{d}^2y}{\mathrm{d}x^2} + 3\frac{\mathrm{d}y}{\mathrm{d}x}\frac{\left(\mathrm{d}^2x\right)^2}{\mathrm{d}x^4} \tag{17}$$

Because the expanded notation for the second and higher derivatives is much more verbose than the first derivative, it is often useful for clarity and succinctness to write derivatives using a slight modification of Arbogast's D notation (see [9]) for the total derivative instead of writing it as algebraic differentials. Here, we will also be subscripting the D with the variable with which the derivative is being taken with respect to and supplying in the superscript the number of derivatives we are taking. Therefore, where Arbogast would write simply D, this notation would be written as D_x^1.

Below is the second and third derivative of y with respect to x written using both the enhanced Arbogast notation and as a ratio of differentials.

$$D_x^2y = \frac{\mathrm{d}^2y}{\mathrm{d}x^2} - \frac{\mathrm{d}y}{\mathrm{d}x}\frac{\mathrm{d}^2x}{\mathrm{d}x^2} \tag{18}$$

$$D_x^3y = \frac{\mathrm{d}^3y}{\mathrm{d}x^3} - \frac{\mathrm{d}y}{\mathrm{d}x}\frac{\mathrm{d}^3x}{\mathrm{d}x^3} - 3\frac{\mathrm{d}^2x}{\mathrm{d}x^2}\frac{\mathrm{d}^2y}{\mathrm{d}x^2} + 3\frac{\mathrm{d}y}{\mathrm{d}x}\frac{\left(\mathrm{d}^2x\right)^2}{\mathrm{d}x^4} \tag{19}$$

This gets even more important as the number of derivatives increases. Each one is more unwieldy than the previous one. However, each level can be converted to differential notation as follows:

$$D_x^ny = \frac{\mathrm{d}\left(D_x^{n-1}y\right)}{\mathrm{d}x} \tag{20}$$

The advantage of Arbogast's notation over Lagrangian notation are that this modification of Arbogast's notation clearly specifies both the variable/expression whose derivative is being taken and the variable/expression it is being taken with respect to.

Therefore, when a compact representation of higher order derivatives is needed, this paper will use Arbogast's notation for its clarity and succinctness. This notation can be easily expanded to its differentials when necessary for manipulation.

6. Extending the partial derivative's algebraic manipulability

The derivative gives the rate at which a function f changes when x is increased. But what if f depends on both x and y? Imagine a hill where f is the distance above sea level, x is the distance east from the origin, and y is the distance north from the origin. To find how f is changing, a direction to measure the slope must be picked. Along the direction straight east, only x is changing while y stays constant. This slope is the partial derivative of f with respect to x, denoted $\frac{\partial f}{\partial x}$, the change in f over the change in x when x is the only variable allowed to change ([5], pp. 940–941). A derivative where all the independent variables are allowed to change is called a total derivative, like the two-dimensional derivative $\frac{\mathrm{d}y}{\mathrm{d}x}$. This partial derivative can be formally defined using limits or using hyperreals.

Using limits, the partial derivative of $f(x, y)$ at the point (a, b) with respect to x is $\lim_{h \to 0} \frac{f(a+h,\ b)-f((a,\ b)}{h}$ ([5], p. 941). Likewise, the partial derivative of $f(x, y)$ with respect to x is $\lim_{h \to 0} \frac{f(x+h,\ y)-f(x,\ y)}{h}$. For more than two variables, the partial derivative of $f(x_1, x_2, \ldots)$ with respect to x_1 is

$$\frac{\partial f}{\partial x_1} = \lim_{h \to 0} \frac{f(x_1 + h, x_2, \ldots) - f(x_1, x_2, \ldots)}{h} \tag{21}$$

Like the with the total derivative, using limits to define the partial derivative means the change in f and the change in x are not defined separately and must be kept together. Using hyperreals, the partial derivative of f with respect to x_1 is

$$\frac{\partial f}{\partial x_1} = \frac{f(x_1 + dx_1, x_2, \ldots) - f(x_1, x_2, \ldots)}{dx_1} \tag{22}$$

Also, dx_1 can equal ∂x_1 assuming both of them denote the smallest change in x_1 possible. This is not an equation in the real numbers; it is an equation in the hyperreals.

Both the numerator and denominator of $\frac{\partial f}{\partial x_1}$ have meaning on their own, and they both are specific hyperreals. So it should be possible to separate the fraction without problems.

However, the current notation for ∂f does not distinguish between the change in f when x_1 is allowed to change and the change in f when another variable, say x_2, is allowed to change. In other words, the ∂f in $\frac{\partial f}{\partial x_1}$ is a different hyperreal from the ∂f in $\frac{\partial f}{\partial x_2}$, even though they both use the exact same symbol. This can cause problems if the notation is taken seriously (see the contradiction noted in Section 2). Adding more information to the notation resolves this issue.

The notation for the partial derivative should be changed from $\frac{\partial f}{\partial x}$ to $\frac{\partial(f,\ x)}{dx}$ in order to preserve the information in the numerator when the fraction is separated.

This makes it clear that ∂ is an operator that takes as an argument not only f but also the choice of which variable to vary. The function that ∂ acts on, in this case f, is the first argument of ∂ and every argument after the first is a variable allowed to change. This can lead to expressions like $\partial(f, x, y)$, the change in f when both x and y are allowed to vary.

Using this notation, $\frac{df}{dt}$ equals $\frac{\partial(f,\ x)}{dt} + \frac{\partial(f,\ y)}{dt}$, not $\frac{\partial f}{dt} + \frac{\partial f}{dt}$. The contradictions are resolved, and the partial derivative fraction can be separated. The numerator and denominator can be moved around just like any other algebraic expression, keeping in mind both of them are hyperreals, so technically any operations on them should be hyperreal operations.

Because the new notation can be algebraically manipulated without contradictions, it makes possible new equations where infinitesimals are not confined to ratios. For instance, the resolved contradiction proof gave the equation $df = \partial(f, x) + \partial(f, y)$. This is reminiscent of one of the conditions for differentiability, $\Delta f = f_x(a, b)\Delta x + f_y(a, b)\Delta y + \varepsilon_1 \Delta x + \varepsilon_2 \Delta y$, where for fixed a and b, ε_1 and ε_2 are functions that depend only on Δx and Δy, with $\varepsilon_1, \varepsilon_2 \to (0, 0)$ as $(\Delta x, \Delta y) \to (0, 0)$ ([5], p. 947).

Besides simplifying old equations, with the new notation it is possible to consider individual partial changes when building equations, just like considering individual total changes.

The new notation can also denote expressions like $\partial(f, x_1, x_2)$, the change in $f(x_1, x_2, x_3)$ when x_1 and x_2 are allowed to vary, but x_3 must stay constant. With the current notation ∂f, dealing with these situations is clumsy at best.

$\partial(f, x_1)$ is an infinitesimal with meaning on its own. It can be defined analogously to Eq. 12:

$$\partial(f, x_1) = f(x_1 + \mathrm{d}x_1, x_2 \ldots) - f(x_1, x_2 \ldots) \tag{23}$$

The total differential of f is usually defined as the combination of all of the changes in f depending on each variable. Typically, the total differential of a multivariate function is found using the sum of its partial *derivatives* multiplied by their respective differentials.

$$\mathrm{d}f(x_1, x_2 \ldots) = \frac{\partial f}{\partial x_1}\mathrm{d}x_1 + \frac{\partial f}{\partial x_2}\mathrm{d}x_2 + \ldots \tag{24}$$

Using the new definition of the partial differential, we can rewrite the formula much more straightforwardly, where the total differential is simply a sum of its partial differentials.

$$\mathrm{d}f(x_1, x_2 \ldots) = \partial(f, x_1) + \partial(f, x_2) + \ldots \tag{25}$$

7. Building differential formulas

Using the notation established in this paper, we can build standard calculus formulas in a clear, algebraic manner. The notation and the formulas will flow directly from the basic truths of calculus and the algebraic reasoning of differentials.

7.1 The inverse function theorem for second derivatives

The standard inverse function theorem simply states that $\frac{\mathrm{d}x}{\mathrm{d}y} = \frac{1}{\frac{\mathrm{d}y}{\mathrm{d}x}}$. In other words, as implied by the algebraic arrangement of its terms, the derivative of x with respect to y is simply the inverse of the derivative of y with respect to x. Using the hyperreal understanding of derivatives allows for a more straightforward way of considering this fact.

More importantly, the new notation for the second derivative likewise allows for a straightforward algebraic construction of an inverse function theorem for the second derivative. Since the second derivative of y with respect to x is $D_x^2 y = \frac{\mathrm{d}^2y}{\mathrm{d}x^2} - \frac{\mathrm{d}y}{\mathrm{d}x}\frac{\mathrm{d}^2x}{\mathrm{d}x^2}$, then the second derivative of x with respect to y will likewise be $D_y^2 x = \frac{\mathrm{d}^2x}{\mathrm{d}y^2} - \frac{\mathrm{d}x}{\mathrm{d}y}\frac{\mathrm{d}^2y}{\mathrm{d}y^2}$. Is there a way to construct a formula for converting one to the other? A simple multiplication by $-\left(\frac{\mathrm{d}x}{\mathrm{d}y}\right)^3$ yields

$$-D_x^2 y\left(\frac{\mathrm{d}x}{\mathrm{d}y}\right)^3 = \frac{\mathrm{d}^2x}{\mathrm{d}y^2} - \frac{\mathrm{d}^2y}{\mathrm{d}y^2}\frac{\mathrm{d}x}{\mathrm{d}y}$$

Here, $\frac{\mathrm{d}x}{\mathrm{d}y}$ can be trivially recognized as $\frac{1}{D_x^1 y}$, and the right-hand side of the equation can be recognized as $D_y^2 x$. Therefore, this can be rewritten as

$$-D_{x}^{2}y\left(\frac{1}{D_{x}^{1}y}\right)^{3}=D_{y}^{2}x \tag{26}$$

which is the inverse function theorem for the second derivative.

7.2 The chain rule for the second derivative

The chain rule for the second derivative can also be easily derived from the new notation. Starting with the notation for the second derivative of y with respect to x, we can look at the transformations needed to generate a second derivative of y with respect to t. We will start by multiplying by $\frac{\mathrm{d}x^2}{\mathrm{d}t^2}$ in order to match the leading term to what is needed for the final result.

$$D_{x}^{2}y=\frac{\mathrm{d}^{2}y}{\mathrm{d}x^{2}}-\frac{\mathrm{d}y}{\mathrm{d}x}\frac{\mathrm{d}^{2}x}{\mathrm{d}x^{2}} \tag{27}$$

$$D_{x}^{2}y\left(D_{t}^{1}x\right)^{2}=\frac{\mathrm{d}^{2}y}{\mathrm{d}x^{2}}\frac{\mathrm{d}x^{2}}{\mathrm{d}t^{2}}-\frac{\mathrm{d}y}{\mathrm{d}x}\frac{\mathrm{d}^{2}x}{\mathrm{d}x^{2}}\frac{\mathrm{d}x^{2}}{\mathrm{d}t^{2}} \tag{28}$$

$$D_{x}^{2}y\left(D_{t}^{1}x\right)^{2}=\frac{\mathrm{d}^{2}y}{\mathrm{d}t^{2}}-\frac{\mathrm{d}y}{\mathrm{d}x}\frac{\mathrm{d}^{2}x}{\mathrm{d}t^{2}} \tag{29}$$

In (29) we see that the leading term is what we want, but the second term is problematic. However, it looks a little like the leading term of the second derivative of x with respect to t multiplied by the first derivative of y with respect to t. Adding that combination to our existing result will yield the desired effect.

$$\left(D_{x}^{2}y\right)\left(D_{t}^{1}x\right)^{2}+\left(D_{x}^{1}y\right)\left(D_{t}^{2}x\right)=\frac{\mathrm{d}^{2}y}{\mathrm{d}t^{2}}-\frac{\mathrm{d}y}{\mathrm{d}x}\frac{\mathrm{d}^{2}x}{\mathrm{d}t^{2}}+\frac{\mathrm{d}y}{\mathrm{d}x}\frac{\mathrm{d}^{2}x}{\mathrm{d}t^{2}}-\frac{\mathrm{d}y}{\mathrm{d}x}\frac{\mathrm{d}x}{\mathrm{d}t}\frac{\mathrm{d}^{2}t}{\mathrm{d}t^{2}} \tag{30}$$

$$\left(D_{x}^{2}y\right)\left(D_{t}^{1}x\right)^{2}+\left(D_{x}^{1}y\right)\left(D_{t}^{2}x\right)=\frac{\mathrm{d}^{2}y}{\mathrm{d}t^{2}}-\frac{\mathrm{d}y}{\mathrm{d}t}\frac{\mathrm{d}^{2}t}{\mathrm{d}t^{2}} \tag{31}$$

As is evident, the right-hand side is the desired result—the second derivative of y with respect to t.

7.3 The chain rule for multivariate derivatives

Building the chain rule for multivariate derivatives is even more straightforward. Consider a function $f(x, y)$ where x and y are both functions of t. As noted in (25), The total change in f, $\mathrm{d}f$, has two parts: the change due to x changing and the change due to y changing. So,

$$\mathrm{d}f=\partial(f,x)+\partial(f,y) \tag{32}$$

Dividing both sides by $\mathrm{d}t$,

$$\frac{\mathrm{d}f}{\mathrm{d}t}=\frac{\partial(f,x)}{\mathrm{d}t}+\frac{\partial(f,y)}{\mathrm{d}t} \tag{33}$$

This is a valid equation, but it is difficult to calculate a value like $\frac{\partial(f, x)}{dt}$ directly. To make it easier to work with, we can multiply the first term by $\frac{dx}{dx}$ and the second by $\frac{dy}{dy}$: [3]

$$\frac{df}{dt} = \frac{\partial(f, x)}{dt} \cdot \frac{dx}{dx} + \frac{\partial(f, y)}{dt} \cdot \frac{dy}{dy} \tag{34}$$

$$= \frac{\partial(f, x)}{dx} \cdot \frac{dx}{dt} + \frac{\partial(f, y)}{dy} \cdot \frac{dy}{dt} \tag{35}$$

This is the standard chain rule for multivariate derivatives.

8. Conclusion

While treating derivatives as ratios of differentials has been long viewed as problematic, small changes in both the understanding and notation of derivatives straightforwardly leads to algebraically manipulable differentials for both total and partial differentials. These differentials provide a more straightforward basis for both doing calculus operations and deriving standard calculus rules. It eliminates exceptions and memorized formulas in favor of simply using algebra with differentials.

Our hope is that the flexibility and freedom of manipulability that this notation allows will both reduce the cognitive load for learning to use differential operators as well as allow for easier exploration of possibilities for practitioners.

Acknowledgements

The authors wish to thank Dr. Enrique Valderrama for his comments on early drafts of this manuscript.

[3] Technically, both $\frac{dx}{dx}$ and $\frac{dy}{dy}$ equal [1], not 1. But, since this is an equation in the hyperreals (with hyperreal multiplication), multiplying by the hyperreal multiplication identity does not change the value of the right side of the equation.

Author details

Maria Isabelle Fite[1†] and Jonathan Bartlett[2*†]

1 University of Tulsa, Tulsa, Oklahoma, United States

2 The Blyth Institute, Tulsa, Oklahoma, United States

*Address all correspondence to: jonathan.bartlett@blythinstitute.org

† These authors contributed equally.

References

[1] Johnson WP. The curious history of fa'a di Bruno's formula. The American Mathematical Monthly. 2002;**109**(3): 217-234. DOI: 10.1080/00029890. 2002.11919857

[2] Newton I. The Method of Fluxions and Infinite Series; with its Application to the Geometry of Curve-Lines, (Translated by John Colson). London: Henry Woodfall and John Nourse; 1736

[3] Bell JL. Continuity and Infinitesimals. The Stanford Encyclopedia of Philosophy. 2022 ed. Stanford, CA: Spring, The Metaphysics Research LabPhilosophy Department Stanford University; 2022

[4] Berkeley G. The Analyst: A Discourse Addressed to an Infidel Mathematician. London: J. and R. Tonson and S. Draper; 1734

[5] Briggs W, Cochran L, Gillett B, Schulz E. Calculus: Early Transcendentals. 3rd ed. New York: Pearson Education; 2019

[6] Herrmann RA. Nonstandard analysis: A simplified approach. arXiv. 2010; **math/0310351v6**: 1-82

[7] Bartlett J, Gaastra L, Nemati D. Hyperreal numbers for infinite divergent series. Communications of the Blyth Institute. 2020;**2**(1):7-15. DOI: 10.33014/ issn.2640-5652.2.1.bartlett-et-al.1

[8] Bartlett J, Khurshudyan AZ. Extending the algebraic manipulability of differentials. Dynamics of Continuous, Discrete and Impulsive Systems Series A: Mathematical Analysis. 2019;**26**:217-230

[9] Cajori F. A History of Mathematical Notations. Vol. II. Chicago: Open Court Publishing; 1929

Chapter 5

Stabilization of a Quantum Equation under Boundary Connections with an Elastic Wave Equation

Hanni Dridi

Abstract

The stability of coupled PDE systems is one of the most important topic because it covers realistic modeling of the most important physical phenomena. In fact, the stabilization of the energy of partial differential equations has been the main goal in solving many structural or microstructural dynamics problems. In this chapter, we investigate the stability of the Schrödinger-like quantum equation in interaction with the mechanical wave equation caused by the vibration of the Euler–Bernoulli beam, to effect stabilization, viscoelastic Kelvin-Voigt dampers are used through weak boundary connection. Firstly, we show that the system is well-posed via the semigroup approach. Then with spectral analysis, it is shown that the system operator of the closed-loop system is not of compact resolvent and the spectrum consists of three branches. Finally, the Riesz basis property and exponential stability of the system are concluded via comparison method in the Riesz basis approach.

Keywords: wave equation, exponential stability, Riesz basis approach, C_0–semigroup, spectral analysis

1. Introduction

There are many coupled systems that have been addressed in the literature, and we can hint here that coupling may be through the association of PDEs with coefficients or via boundary conditions of PDEs. The coupling may be strong or weak as the characteristic is determined based on the results obtained after studying the stability or control. We can divide the coupled systems according to the coupling form. Firstly, the parabolic-hyperbolic coupled systems, such as heat wave system, that arise from the interaction of the fluid structure. See works [1, 2] where stability and control systems are analyzed. Secondly, we can refer heat-beam system through works [3, 4] where the researchers used an effective method for stabilization of the system. Thirdly, in the heat-Schrödinger system, the heat dynamic controller was applied for stabilization and Gevrey regularity property in the paper [5]. Finally, in the case of thermoelastic systems, the exponential stability and Riesz basis property of the coupled heat equation and elastic structure were discussed in reference [6]. The exponential stability of thermoplastic systems with microtemperature in reference [7], for the linear beam system coupled with thermal effect, we refer to the works [8–12]. For the nonlinear beam system with thermal effect, see reference [13].

From general result related to the previously mentioned research works, we can conclude that the heat equation plays the role of dynamic boundary feedback controller of the hyperbolic PDE. Also, for the interconnected system of Euler–Bernoulli beam and heat equation with boundary weak connections where the heat is the dynamic boundary controller to the whole system, which means that this subsystem can be presented as a controller for other subsystems.

Euler–Bernoulli beam equation with boundary energy dissipation is analyzed in the work [14], the problem is given as follows:

$$\begin{cases} \rho y_{tt} + EIy_{xxxx} = 0, & 0<x<1, \\ y(0,t) = y_x(0,t) = 0 & \\ -EIy_{xxx}(1,t) = -k_1^2 y_t(1,t), & k_1 \in \mathbb{R}, \\ -EIy_{xx}(1,t) = k_2^2 y_{xt}(1,t), & k_2 \in \mathbb{R}, \\ y(x,0) = y_0(x) \quad y_t(x,0) = y_1(x), & 0 \le x \le 1, \end{cases} \tag{1}$$

where ρ denotes the mass density per unit length, EI is the flexural rigidity coefficient. The authors extract some estimates of the resolvent operator on the imaginary axis by applying Huang's[1] theorem to establish an exponential decay result.

For the asymptotic behavior of the wave equation, we introduce the following problem:

$$\begin{cases} \dfrac{\partial^2 w}{\partial t^2} - \Delta w = 0 \quad \text{in} & \Omega \times (0,\infty), \\ w(x,t) = 0 \quad \text{on} & \Gamma_0 \times [0,\infty), \\ \dfrac{\partial w}{\partial \nu} + a(x)\dfrac{\partial w}{\partial t} = 0 \quad \text{on} & \Gamma_1 \times (0,\infty), \end{cases} \tag{2}$$

where ν is the unit normal of Γ pointing toward exterior of Ω. The function $a \in C^1(\overline{\Gamma_1})$ with $a(x) \ge a_0 > 0$ on Γ_1. Problem (2) has been treated by Lagnese in [17], he used a multiplier method[2] and proved that the energy decay rate is obtained for solutions of wave type equations in a bounded region in $\mathbb{R}^n$ $(n \ge 2)$ whose boundary consists partly of a nontrapping reflecting surface and partly of an energy absorbing surface. We can express this result, as follows:

$$E(t) \le f(t)E(0), \qquad t \ge 0, \tag{3}$$

[1] Huang [15] introduced a frequency domain method to study the exponential decay of such stability problems.

[2] The energy multiplier method [16, 17] has been successfully applied to establish exponential stability, which is a very desirable property for elastic systems.

with energy defined by

$$E(t) = \frac{1}{2}\left(\|w_t\|^2_{L^2(\Omega)} + \|\nabla w\|^2_{L^2(\Omega)}\right). \tag{4}$$

The decay rate of solutions is a function $f(t)$ satisfying $f(t) \to 0$ as $t \to \infty$. However, there are difficulties with some boundary condition problems, which makes the energy multiplier method ineffective in proving the exponential stability property.

Wazwaz [18], used the variational iteration method[3] for the study of both linear and nonlinear Schrödinger equations, these problem is governed by the following equations:

$$\begin{cases} u_t + iu_{xx} = 0, \\ u(x, 0) = f(x), \qquad i^2 = -1 \end{cases} \tag{5}$$

and

$$\begin{cases} iu_t + u_{xx} + \gamma |u|^{2r} u = 0, \quad r \geq 1, \\ u(x, 0) = f(x), \quad i^2 = -1. \end{cases} \tag{6}$$

The variational iteration method was used to give rapid convergent successive approximations as well as to treat linear and non-linear problems in a uniform manner.

1.1 Statement of the problem

In this work, we consider stabilization for a Schrödinger equation through a boundary feedback dynamic controller interacted by an Euler–Bernoulli beam equation with Kelvin-Voigt damping[4], the system is described by the following coupled partial differential equations:

$$\begin{cases} \partial_t^2 u + \partial_x^4 u + \beta \partial_x^4 \partial_t u = 0, \quad & 0 < x < 1, t > 0, \\ \partial_t v + i\partial_x^2 v = 0, & 0 < x < 1, t > 0, \end{cases} \tag{7}$$

boundary conditions are given by

$$\begin{cases} u(1, t) = \partial_x u(0, t) = \partial_x^2 u(1, t) = v(1, t) = 0, \quad & t \geq 0, \\ v(0, t) = \alpha \partial_t u(0, t), & t \geq 0, \\ \beta \partial_x^3 \partial_t u(0, t) + \partial_x^3 u(0, t) = -\alpha i \partial_x v(0, t), & t \geq 0, \end{cases} \tag{8}$$

the problem is associated with the following initial conditions:

$$u(x, 0) = u_0(x), \partial_t u(x, 0) = u_1(x), v(x, 0) = v_0(x), \qquad 0 \leq x \leq 1. \tag{9}$$

---

[3] The variational iteration method is established by He in [19, 20] is thoroughly used by many researchers to handle linear and nonlinear models.

[4] Kelvin-Voigt is one of the most important types of damping and has been used in many works, see for example, [10, 21].

1.2 Energy space

Initial condition (9) is in the following phase space:

$$\mathcal{H} = H^2_*(0,1) \times L^2(0,1) \times L^2(0,1), \tag{10}$$

where

$$H^2_*(0,1) = \{s | s \in H^2(0,1), \partial_x s(0) = s(1) = 0\}.$$

1.3 Energies

The energy is the sum of the potential energy and the kinetic energy, given by

$$E(t) = \frac{1}{2}\left(\|u_t\|^2_{L^2(0,1)} + \|\partial^2_x u\|^2_{L^2(0,1)} + \|v\|^2_{L^2(0,1)}\right). \tag{11}$$

Then, we have

$$\frac{d}{dt}E(t) = -\beta\|\partial^2_x \partial_t u\|^2_{L^2(0,1)}. \tag{12}$$

It is clear that $E(t)$ is nonincreasing with time.

1.4 Remark

1. The energy dissipation is related to the wave equation, that is, there are no explicit terms for a part of the Schrödinger subsystem.

2. We note that the weakness of the boundary connections for problems (7)–(9) lead to a complicated problem in stability analysis.

3. If we take the β coefficient equal to zero in Eq. (12), the system becomes conservative.

1.5 Notations

1. $\langle \cdot, \cdot \rangle_{L^2(0,1)}$ is the $L^2(0,1)-$inner product and $\|\cdot\|_{L^2(0,1)}$ is the $L^2(0,1)-$norm.

2. The symbols $\Re(s)$ and (s) indicate the real part and the imaginary of a complex number s.

3. $(s)^T$ represents the transposed vector of (s).

2. Well-posedness

2.1 Setting of the semigroup

Setting $z = (u, \partial_t u = w, v)^T$. Then, we introduce the norm in the Hilbert space $\mathcal{H}$ as follows:

$$\begin{aligned} \|z\|_{\mathcal{H}}^2 &= \|u_t\|_{L^2(0,1)}^2 + \left\|\partial_x^2 u\right\|_{L^2(0,1)}^2 + \|v\|_{L^2(0,1)}^2 \\ &= 2E(t), \end{aligned} \tag{13}$$

for $z_1, z_2 \in \mathcal{H}$, the norm (13) is induced by the following inner product

$$\langle z_1, z_2 \rangle_{L^2(0,1)} = \langle w_1, w_2 \rangle_{L^2(0,1)} + \left\langle \partial_x^2 u_1, \partial_x^2 u_2 \right\rangle_{L^2(0,1)} + \langle v_1, v_2 \rangle_{L^2(0,1)}. \tag{14}$$

System (7) can be written as an abstract Cauchy problem in the phase space (10) as follows:

$$\begin{cases} \dfrac{d}{dt} z = \mathcal{A}z, t > 0, \\ z(0) = z_0. \end{cases} \tag{15}$$

The solution at time $t > 0$ to problem (15) can be written as:

$$z(t) = S(t)z_0 = e^{t\mathcal{A}} z_0,$$

where the operator $\mathcal{A} : \mathcal{D}(\mathcal{A}) \subset \mathcal{H} \to \mathcal{H}$ is given by

$$\mathcal{A}z = \begin{pmatrix} w \\ -\partial_x^2 \left(\partial_x^2 u + \beta \partial_x^2 w \right) \\ -i \partial_x^2 v \end{pmatrix}, \tag{16}$$

with domain

$$\mathcal{D}(\mathcal{A}) = \left\{ z \in \mathcal{H}, \mathcal{A}z \in \mathcal{H} \middle| \begin{array}{c} \partial_x^2 u + \beta \partial_x^2 w \in H^2(0,1), \\ u(1) = \partial_x u(0) = \partial_x^2 u(1) = v(1) = 0, \\ v(0) = \alpha w(0), \\ \beta \partial_x^3 w(0) + \partial_x^3 u(0) = -\alpha i \partial_x v(0), \end{array} \right\}, \tag{17}$$

Theorem 1.1: Let $\mathcal{A}$ defined by (16). Then, $\mathcal{A}^{-1}$ exists and $\mathcal{A}$ generates a C_0-semigroup of contractions on $\mathcal{H}$.

Proof: We use the semigroup method, we shall show that:

1. The operator $\mathcal{A}$ is dissipative.

2. The operator $I_d - \mathcal{A}$ is onto (I_d is the identity operator).

For the proof of (1). Firstly, we have $\mathcal{D}(\mathcal{A})$ is dense in $\mathcal{H}$, that is,

$$\overline{\mathcal{D}(\mathcal{A})} = \mathcal{H}. \tag{18}$$

Secondly, by applying the scalar product in the Hilbert space $\mathcal{H}$, we obtain

$$\begin{aligned} \langle \mathcal{A}z, z \rangle_{\mathcal{H}} &= \left\langle \partial_x^2 w, \partial_x^2 \overline{u} \right\rangle_{L^2(0,1)} - \left\langle \partial_x^2 \left(\partial_x^2 u + \beta \partial_x^2 w \right), \overline{w} \right\rangle_{L^2(0,1)} - \left\langle i \partial_x^2 v, \overline{v} \right\rangle_{L^2(0,1)} \\ &= \left\langle \partial_x^2 w, \partial_x^2 \overline{u} \right\rangle_{L^2(0,1)} + \left(\partial_x^3 u(0) + \beta \partial_x^3 w(0) \right) \overline{w}(0) \\ &\quad + i \partial_x v(0) \overline{v}(0) + \langle i \partial_x v, \partial_x \overline{v} \rangle_{L^2(0,1)} - \left\langle \partial_x^2 u + \beta \partial_x^2 w, \partial_x^2 \overline{w} \right\rangle_{L^2(0,1)}. \end{aligned} \tag{19}$$

By using boundary conditions (8), we get

$$\Re\langle \mathcal{A}z, z\rangle_{\mathcal{H}} = -\beta\left\|\partial_x^2 w\right\|_{L^2(0,1)}^2 \leq 0. \tag{20}$$

Then, the density property (18) and inequality (20) show that $\mathcal{A}$ is dissipative. For the proof of (2), we shall solve the equation

$$\mathcal{A}z = F$$

for any $F = (f_1, f_2, f_3)^T \in \mathcal{H}$, we can express the equation as follows:

$$\begin{cases} w = f_1, \\ \partial_x^2\left(\partial_x^2 u + \beta\partial_x^2 w\right) = -f_2, \\ i\partial_x^2 v = -f_3 \end{cases} \tag{21}$$

By using the first equation of (21), we get

$$\begin{cases} \partial_x^4 u = -f_2 + \beta\partial_x^4 f_1, \\ \partial_x^2 v = if_3. \end{cases} \tag{22}$$

We solve the following equation for the function v,

$$\begin{cases} \partial_x^2 v = if_3, \\ v(1) = 0, \quad v(0) = \alpha f_1(0), \end{cases} \tag{23}$$

to obtain

$$\begin{cases} v = \partial_x v(0)x + i\int_0^x (x-y)f_3(y)dy + \alpha f_1(0), \\ \partial_x v(0) = -i\int_0^1 (1-y)f_3(y)dy - \alpha f_1(0). \end{cases} \tag{24}$$

For u, we solve

$$\begin{cases} \partial_x^4 u = -f_2 + \beta\partial_x^4 f_1, \\ u(1) = \partial_x u(0) = \partial_x^2 u(1) = 0, \\ \beta\partial_x^3 w + \partial_x^3 u(0) = -i\alpha\partial_x v(0), \end{cases} \tag{25}$$

to obtain

$$\begin{cases} u = -\int_0^x (1-x)g(y)dy - \int_x^1 (1-y)g(y)dy, \\ g(x) = \beta\left(\partial_x^2 f_1(1) - \partial_x^2 f_1(x)\right) + \int_0^x (1-x)f_2(y)dy \\ \qquad + \int_x^1 (1-y)f_2(y)dy + i\alpha\partial_x v(0)(1-x). \end{cases} \tag{26}$$

Eqs. (24) and (26) give a unique $z \in \mathcal{D}(\mathcal{A})$ satisfying $\mathcal{A}z = F$.
It is easy to check that $\mathcal{A}^{-1}$ is bounded, that is,

$$0 \in \rho(\mathcal{A}).$$

Therefore, the operator $\mathcal{A}$ generates a C_0-semigroup of contractions on $\mathcal{H}$ by the Lumer–Philips theorem [22].

3. Spectral analysis

We consider the following eigenvalue problem for the system operator $\mathcal{A}$. Let $\mathcal{A}z = \lambda z$. Then, we have

$$\begin{cases} w = \lambda u, \\ \partial_x^2\left(\partial_x^2 u + \beta \partial_x^2 w\right) = -\lambda w, \\ \partial_x^2 v = i\lambda v, \\ u(1) = \partial_x u(0) = \partial_x^2 u(1) = v(1) = 0, \\ \alpha\lambda u(0) = v(0), \\ (1+\beta\lambda)\partial_x^3 u(0) = -i\alpha\partial_x v(0). \end{cases} \tag{27}$$

The first and second equations of system (27) give the following system

$$\begin{cases} (1+\beta\lambda)\partial_x^4 u + \lambda^2 u = 0, \\ \partial_x^2 v = i\lambda v, \\ u(1) = \partial_x u(0) = \partial_x^2 u(1) = v(1) = 0, \\ \alpha\lambda u(0) = v(0), \\ (1+\beta\lambda)\partial_x^3 u(0) = -i\alpha\partial_x v(0). \end{cases} \tag{28}$$

Lemma
For any $\lambda \in \sigma_p(\mathcal{A})$, it holds

$$\Re(\lambda) < 0. \tag{29}$$

Proof: By Theorem 1.1, we have $\Re(\lambda) \le 0$.[5] Letting $0 \ne \lambda \in \sigma_p(\mathcal{A})$ with $\Re(\lambda) = 0$ and $z \in \mathcal{D}(\mathcal{A})$ satisfying

$$\mathcal{A}z = \lambda z. \tag{30}$$

By using inequality 20, it follows that

$$0 = \Re(\lambda)\|z\|_{\mathcal{H}}^2 = \Re\langle \mathcal{A}z, z\rangle_{\mathcal{H}} = -\beta\left\|\partial_x^2 w\right\|_{L^2(0,1)}^2. \tag{31}$$

From Eq. (31) and boundary conditions $(28)_3$, we have $w = 0$.

---

[5] $\mathcal{A}$ is dissipative $\Rightarrow \Re(\lambda) \le 0, \ \forall \lambda \in \sigma_p(\mathcal{A})$.

From $(27)_1$ we have $u = 0$. Moreover, Eq. (30) gives

$$\begin{cases} \partial_x^2 v = i\lambda v, \\ v(0) = v(1) = \partial_x v(0) = 0. \end{cases} \tag{32}$$

It is easy to check that the above equation has only a trivial null solution $v = 0$. Hence, $z = 0$, and all the points that are located on the imaginary axis are not eigenvalues of $\mathcal{A}$. Then the proof is completed.

Setting $\lambda = \rho^2$ in (28), when $1 + \beta\rho^2 \neq 0$, we obtain

$$\begin{cases} \partial_x^4 u = \dfrac{-\rho^4}{1+\beta\rho^2} u, \\ \partial_x^2 v = i\rho^2 v, \\ u(1) = \partial_x u(0) = \partial_x^2 u(1) = v(1) = 0, \\ \alpha\rho^2 u(0) = v(0), \\ (1+\beta\rho^2)\partial_x^3 u(0) = -i\alpha\partial_x v(0). \end{cases} \tag{33}$$

Let

$$a = \sqrt[4]{\frac{-\lambda^2}{1+\beta\lambda}}.$$

Then, the general solution of system (33) can be expressed as follows:

$$\begin{aligned} u &= c_1 \exp(ax) + c_2 \exp(-ax) + c_3 \exp(iax) + c_4 \exp(-iax), \\ v &= d_1 \exp\left(\sqrt{i\rho} x\right) + d_2 \exp\left(-\sqrt{i\rho} x\right). \end{aligned} \tag{34}$$

By the boundary conditions of (33), we obtain that the constants $c_1, \cdots, c_4$ and d_1, d_2 are not identical to zero if and only if $\det(X) = 0$, where

$$X = \begin{pmatrix} e^a & e^{-a} & e^{ia} & e^{-ia} & 0 & 0 \\ a^2e^a & a^2e^{-a} & -a^2e^{ia} & -a^2e^{-ia} & 0 & 0 \\ a & -a & ia & -ia & 0 & 0 \\ 0 & 0 & 0 & 0 & e^{\sqrt{i\rho}} & -e^{\sqrt{i\rho}} \\ \alpha\rho^2 & \alpha\rho^2 & \alpha\rho^2 & \alpha\rho^2 & -1 & -1 \\ a^3 & -a^3 & -ia^3 & ia^3 & \dfrac{i\sqrt{i}\alpha\rho}{\beta\rho^2+1} & -\dfrac{i\sqrt{i}\alpha\rho}{\beta\rho^2+1} \end{pmatrix}, \tag{35}$$

by using boundary conditions (8), we get

$$c_2 = -e^{2a}c_1, \quad c_4 = -e^{2ia}c_3, \quad d_2 = -e^{2\sqrt{i\rho}}d_1.$$

Then, the solution can be expressed by

$$u = c_1\left(e^{ax} - e^{a(2-x)}\right) + c_3\left(e^{iax} - e^{ia(2-x)}\right), \quad v = d_1\left(e^{\sqrt{i\rho}x} - e^{\sqrt{i\rho}(2-x)}\right),$$

where c_1, c_3, d_1 are determined by the remaining three boundary conditions of (36) that $\det(X) = 0$ if and only if $\det(\tilde{X}) = 0$, where

$$\tilde{X} = \begin{pmatrix} 1+e^{2a} & i+ie^{2ia} & 0 \\ (1-e^{2a})\alpha\rho^2 & (1-e^{2ia})\alpha\rho^2 & -1+e^{2\sqrt{i\rho}} \\ a^3(1+e^{2a}) & -ia^3(1+e^{2ia}) & \frac{i\sqrt{i\alpha\rho}}{\beta\rho^2+1}\left(1+e^{2\sqrt{i\rho}}\right) \end{pmatrix}. \tag{36}$$

We recall the result of Lemma (29) and in light of this, we know that all eigenvalues have negative real parts. Thus, we only consider those λ that lie in the second and third quadrants of the complex plane:

$$S := \left\{\rho \in \mathbb{C} | \frac{\pi}{4} \le \arg\rho \le \frac{3\pi}{4}\right\}.$$

Denote the region $S := S_1 \cup S_2 \cup S_3$ such that

$$S_1 = \left\{\rho \in \mathbb{C} | \frac{\pi}{4} \le \arg\rho \le \frac{3\pi}{8}\right\},$$
$$S_2 = \left\{\rho \in \mathbb{C} | \frac{3\pi}{8} \le \arg\rho \le \frac{5\pi}{8}\right\},$$
$$S_3 = \left\{\rho \in \mathbb{C} | \frac{5\pi}{8} \le \arg\rho \le \frac{3\pi}{4}\right\},$$

the following theorem gives asymptotic distributions of the eigenvalues in S_1, S_2, and S_3.

Theorem 1.2: The eigenvalues of $\mathcal{A}$ have two families:

$$\sigma_p(\mathcal{A}) = \{\lambda_{1n}, n \in \mathbb{N}\} \cup \{\lambda_{2n}^+, \lambda_{2n}^-, n \in \mathbb{N}\},$$

where

$$\begin{aligned} \lambda_{1n} &= in^2\pi^2 + \frac{\sqrt{2}\alpha^2}{\sqrt[4]{\beta}} e^{\frac{5i\pi}{8}} \sqrt{n\pi} - \frac{\alpha^4}{\sqrt{\beta}} e^{\frac{i\pi}{4}} + O\left(n^{\frac{-1}{2}}\right), \\ \lambda_{2n}^+ &= -\beta\left(n\pi - \frac{\pi}{2}\right)^4 + 4\sqrt{i\beta}\alpha^2\left(n\pi - \frac{\pi}{2}\right)^2 - 2\sqrt{2i}\alpha^4\left(n\pi - \frac{\pi}{2}\right) \\ &\quad + \left(6i\pi\alpha^4 - \frac{2\sqrt{i}\alpha^6}{\sqrt{\beta}}\right) + O\left(\frac{1}{n}\right), \\ \lambda_{2n}^- &= -\frac{1}{\beta} - \frac{1}{\beta^3\left(n\pi - \frac{\pi}{2}\right)^4} + O\left(\frac{1}{n^8}\right). \end{aligned} \tag{37}$$

Therefore, we have

$$\Re(\lambda_{1n}), \Re(\lambda_{2n}^+) \to -\infty, \Re(\lambda_{2n}^-) \to -\frac{1}{\beta} \quad \text{as} \quad n \to \infty.$$

Proof: When $\rho \in S_1$, it has

$$\Re\left(\sqrt{i\rho}\right) = |\rho|\cos\left(\arg\left(\rho + \frac{\pi}{4}\right)\right) \le 0.$$

Since

$$a=\sqrt[4]{\frac{-\lambda^2}{1+\beta\lambda}}=\sqrt[4]{\frac{-\rho^4}{1+\beta\rho^2}}=\frac{\sqrt{i\rho}}{\sqrt[4]{\beta}}+O\left(|\rho|^{-\frac{3}{2}}\right)\quad\text{as}\quad|\rho|\to\infty. \tag{38}$$

Based on estimate (38), we can state that there is a positive constant γ_1 such that

$$-\Re(a)=-\frac{\sqrt{|\rho|}}{\sqrt[4]{\beta}}\cos\left(\arg\left(\sqrt{\rho}+\frac{\pi}{4}\right)\right)\leq-\frac{\sqrt{|\rho|}}{\sqrt[4]{\beta}}\sin\left(\frac{\pi}{16}\right)<-\gamma_1\sqrt{|\rho|},$$
$$\Re(ia)=\frac{\sqrt{|\rho|}}{\sqrt[4]{\beta}}\cos\left(\arg\left(\sqrt{\rho}+\frac{3\pi}{4}\right)\right)\leq-\frac{\sqrt{|\rho|}}{\sqrt[4]{\beta}}\cos\left(\frac{\pi}{8}\right)<-\gamma_1\sqrt{|\rho|}.$$

Therefore, we get the following estimates

$$|e^{-a}|=O\left(e^{-\gamma_1\sqrt{|\rho|}}\right),\ |e^{ia}|=O\left(e^{-\gamma_1\sqrt{|\rho|}}\right),\ |e^{\sqrt{i\rho}}|\leq 1. \tag{39}$$

By multiplying some factors, we make each entry of the $\det(\tilde{X})$ be bounded as $\rho\to\infty$

$$\frac{1}{a^3e^{2a}}\det(\tilde{X})=\begin{vmatrix}1+e^{-2a} & i+ie^{2ia} & 0\\ \alpha e^{-2a}-\alpha & \alpha-\alpha e^{2ia} & -1+e^{2\sqrt{i\rho}}\\ 1+e^{-2a} & -i\left(1+e^{2ia}\right) & \dfrac{i\sqrt{i}\alpha\rho^3}{(\beta\rho^2+1)a^3}\left(1+e^{2\sqrt{i\rho}}\right)\end{vmatrix}. \tag{40}$$

By using the expression of a and ρ, and the Taylor expansion, we obtain

$$\frac{i\sqrt{i}\alpha\rho^3}{(\beta\rho^2+1)a^3}=\frac{\alpha}{\sqrt[4]{\beta}}\sqrt{\frac{1}{\rho}}+O\left(|\rho|^{-\frac{5}{2}}\right). \tag{41}$$

By using Eqs. (41) and (39), we get

$$\begin{aligned}\frac{1}{a^3e^{2a}}\det(\tilde{X})&=\begin{vmatrix}1 & i & 0\\ -\alpha & \alpha & -1+e^{2\sqrt{i\rho}}\\ 1 & -i & \dfrac{\alpha}{\sqrt[4]{\beta}}\sqrt{\dfrac{1}{\rho}}\left(1+e^{2\sqrt{i\rho}}\right)\end{vmatrix}+O\left(|\rho|^{\frac{-5}{2}}\right)\\ &=\left(\frac{(1+i)\alpha^2}{\sqrt[4]{\beta}}\sqrt{\frac{1}{\rho}}-2i\right)+e^{2\sqrt{i\rho}}\left(\frac{(1+i)\alpha^2}{\sqrt[4]{\beta}}\sqrt{\frac{1}{\rho}}+2i\right)\\ &\quad+O\left(|\rho|^{\frac{-5}{2}}\right).\end{aligned} \tag{42}$$

From the previous equality, we can get $\det(\tilde{X})=0$ if and only if

$$e^{2\sqrt{i\rho}}=1-\frac{(1-i)\alpha^2}{\sqrt[4]{\beta}}\sqrt{\frac{1}{\rho}}-\frac{i\alpha^4}{\sqrt{\beta}\rho}+O\left(|\rho|^{\frac{-3}{2}}\right). \tag{43}$$

Suppose

$$2\sqrt{i\rho} = 2n\pi i + O\left(n^{\frac{-1}{2}}\right), \tag{44}$$

where n is a sufficiently large integer. Substituting Eq. (43) into Eq. (42), we arrive at

$$O\left(n^{\frac{-1}{2}}\right) = \frac{(-1)^{\frac{5}{8}}\sqrt{2}\alpha^2}{\sqrt[4]{\beta}\sqrt{n\pi}} - \frac{\sqrt{i}\alpha^4}{n\pi\sqrt{\beta}} + O\left(n^{\frac{-3}{2}}\right). \tag{45}$$

The roots of Eq. (42) have the following asymptotic expressions

$$\rho_{1n} = \sqrt{i}n\pi + \frac{(-1)^{\frac{3}{8}}\alpha^2}{\sqrt[4]{\beta}\sqrt{2n\pi}} - \frac{\alpha^4}{2n\pi\sqrt{\beta}} + O\left(n^{\frac{-3}{2}}\right), n > N_1, \tag{46}$$

where N_1 is a sufficiently large positive integer. By $\lambda = \rho^2$, we have

$$\lambda_{1n} = in^2\pi^2 + \frac{\sqrt{2}\alpha^2}{\sqrt[4]{\beta}} e^{\frac{i\pi}{4}} + O\left(n^{-\frac{1}{2}}\right).$$

By using the value of a given by Eq. (38), we can obtain the expression of a as follows:

$$a_{1n} = \frac{(-1)^{\frac{3}{8}}\sqrt{\pi n}}{\sqrt[4]{\beta}} + O\left(\frac{1}{n}\right). \tag{47}$$

Similarly, when $\rho \in S_2$, it is easier to verify that there exists a $\gamma_2 > 0$ such that

$$\begin{cases} \Re(ia) \leq -\gamma_2\sqrt{|\rho|}, \\ \Re(\sqrt{i\rho}) = |\rho| \cos\left(\arg\left(\rho + \frac{\pi}{4}\right)\right) \leq |\rho| \cos\left(\frac{5\pi}{8}\right). \end{cases}$$

Hence, we get the following estimations

$$|e^{ia}| = O\left(e^{-\gamma_2\sqrt{|\rho|}}\right), \quad |e^{\sqrt{i\rho}}| = O(e^{-\gamma_2|\rho|}),$$

by using Eq. (38), we obtain

$$\arg(a) = \arg\left(\sqrt{i\rho}\right) \in \left(\frac{7\pi}{16}, \frac{9\pi}{16}\right] \quad \text{in} \quad S_2.$$

Thus, the sign of a is different under the two conditions:

$$\arg(\rho) \in \left(\frac{7\pi}{16}, \frac{\pi}{2}\right] \quad \text{and} \quad \arg(\rho) \in \left(\frac{\pi}{2}, \frac{9\pi}{16}\right].$$

Therefore, we conclude that

$$\frac{1}{a^3e^a}\det(\tilde{X}) = \begin{vmatrix} e^{-a}+e^a & i+ie^{2ia} & 0 \\ e^{-a}\alpha - e^a\alpha & \alpha - \alpha e^{2ia} & -1+e^{2\sqrt{i\rho}} \\ e^{-a}+e^a & -i(1+e^{2ia}) & \dfrac{i\sqrt{i}\alpha\rho^3}{(\beta\rho^2+1)a^3}\left(1+e^{2\sqrt{i\rho}}\right) \end{vmatrix}$$

$$= \begin{vmatrix} e^{-a}+e^a & i & 0 \\ e^{-a}\alpha - e^a\alpha & \alpha & -1 \\ e^{-a}+e^a & -i & \dfrac{i\sqrt{i}\alpha\rho^3}{(\beta\rho^2+1)a^3} \end{vmatrix} + O\left(e^{-\gamma_2\sqrt{|\rho|}}\right)$$

$$= e^a\left(\frac{\sqrt{2}a\alpha^2}{\rho} - 2i\right) - e^{-a}\left(\frac{\sqrt{2}ia\alpha^2}{\rho} + 2i\right) + O\left(e^{-\gamma_2\sqrt{|\rho|}}\right).$$

From the previous equality, it is seen that $\det(\tilde{X}) = 0$ if and only if

$$e^a\left(\frac{\sqrt{2}a\alpha^2}{\rho} - 2i\right) - e^{-a}\left(\frac{\sqrt{2}ia\alpha^2}{\rho} + 2i\right) + O\left(e^{-\gamma_2\sqrt{|\rho|}}\right) = 0. \tag{48}$$

By using the expression of a and ρ, we obtain

$$\rho = \sqrt{\beta}a^2 - \frac{1}{2\beta^{\frac{3}{2}}a^2} + O\left(\frac{1}{|a|^4}\right), \tag{49}$$

which shows that $|a|$, $|\rho| \to \infty$ at the same time. Now, substitute the value of ρ given by (48) into equality (47), and we obtain

$$e^a\left(-2i + \frac{\sqrt{2}\alpha^2}{\sqrt{\beta}a} + \frac{\sqrt{2}\alpha^2}{2\beta^{\frac{5}{2}}a^5} + O\left(|a|^{-7}\right)\right) - e^{-a}\left(2i + \frac{i\sqrt{2}\alpha^2}{\sqrt{\beta}a} + \frac{i\sqrt{2}\alpha^2}{2\beta^{\frac{5}{2}}a^5} + O\left(|a|^{-7}\right)\right)$$
$$+O\left(e^{-\gamma_2|a|}\right) = 0.$$

Letting $a = x + iy$, it is easily checked that $\overline{a} = x - iy$ also satisfies the same asymptotic equation above. Hence, we only need to analyze the asymptotic expression of a located in the second quadrant. Given the value of a given by (48), when a is located on the second quadrant, $\Re(-a) \leq 0$ and $|e^{-a}| \leq 1$. Therefore,

$$e^{-2a} = -1 + \frac{(1-i)\alpha^2}{\sqrt{2\beta}a} - \frac{(1-i)\alpha^4}{2a^2\beta} + \frac{(1-i)\alpha^6}{2\sqrt{2}a^3\beta^{\frac{3}{2}}} + O\left(\frac{1}{a^4}\right),$$

and for the quadrant where a is located, we have

$$a_{2n} = i\left(n\pi - \frac{\pi}{2}\right) + \frac{(1+i)\alpha^2}{\sqrt{2\beta}\left(n\pi - \frac{\pi}{2}\right)} - \frac{(1-i)\alpha^4}{2\beta\left(n\pi - \frac{\pi}{2}\right)^2}$$
$$- \frac{(1+i)\alpha^6}{2\sqrt{2}\beta^{\frac{3}{2}}\left(n\pi - \frac{\pi}{2}\right)^3} + O\left(\frac{1}{n^4}\right).$$

Since $a = \sqrt[4]{\frac{\lambda^2}{1+\beta\lambda}}$ or $\lambda^2 - \beta a^4 \lambda - a^4 = 0$, it has

$$\lambda_{2n}^{\pm} = \frac{\beta a^4}{2}\left(1 \pm \sqrt{1 + \frac{4}{\beta^2 a^4}}\right).$$

Using the Taylor expansion, we obtain the expressions of λ_{2n}^{+} and λ_{2n}^{-} given by (37). Moreover, by using $\lambda = \rho^2$, we have the asymptotic expressions of ρ_{2n}^{+} and ρ_{2n}^{-}

$$\begin{cases} \rho_{2n}^{+} = i\sqrt{\beta}\left(n\pi - \frac{\pi}{2}\right)^2 + 2\sqrt{i}\alpha^2 + O(n^{-1}), \\ \rho_{2n}^{-} = \dfrac{i}{\sqrt{\beta}} + \dfrac{i}{2\beta^{\frac{5}{2}}\left(n\pi - \frac{\pi}{2}\right)^4} + O\left(n^{-8}\right). \end{cases} \tag{50}$$

Similarly, in S_3, there exists $\gamma_3 > 0$ such that

$$|e^{a}| = O\left(e^{-\gamma_3\sqrt{|\rho|}}\right), \quad |e^{ia}| = O\left(e^{-\gamma_3\sqrt{|\rho|}}\right), \quad |e^{\sqrt{i}\rho}| = O\left(e^{-\gamma_3|\rho|}\right).$$

It is easy to check that there is no null point of $\det\left(\tilde{X}\right)$, namely, there is no point spectrum in S_3.

According to the conclusion of Theorem 1.2, it is obvious that $-\frac{1}{\beta}$ is an accumulation point of the point spectrum of the operator $\mathcal{A}$. We thus have the following corollary.

Corollary

$$\sigma_c(\mathcal{A}) = -\frac{1}{\beta}. \tag{51}$$

We next analyze the asymptotic expression of eigenfunctions of the operator $\mathcal{A}$.

Theorem 1.3: Let $\sigma_p(\mathcal{A}) = \{\lambda_{1n}, n \in \mathbb{N}\} \cup \left\{\lambda_{2n}^{+}, \lambda_{2n}^{-}, n \in \mathbb{N}\right\}$ be the point spectrum of $\mathcal{A}$. Let $\lambda_{1n} = \rho_{1n}^2$, $\lambda_{2n}^{+} = \left(\rho_{2n}^{+}\right)^2$ and $\lambda_{2n}^{-} = \left(\rho_{2n}^{-}\right)^2$ with ρ_{1n}, ρ_{2n}^{+} and ρ_{2n}^{-} being given by Eqs. (45) and (49), respectively. Then, there are three families of approximated normalized eigenfunctions of $\mathcal{A}$

1. One family $\{z_{1n} = (u_{1n}, \lambda u_{1n}, v_{1n}), n \in \mathbb{N}\}$, where z_{1n} is the eigenfunction of $\mathcal{A}$ corresponding to the eigenvalue λ_{1n}, has the following asymptotic expression:

$$\left(\partial_x^2 u_{1n}, \lambda u_{1n}, v_{1n}\right) = (0, 0, \sin\left[a_n(1-x)\right]) + O_x\left(n^{\frac{-3}{4}}\right), \tag{52}$$

where

$$a_n = n\pi + \frac{(-1)^{\frac{1}{8}}\alpha^2}{\sqrt[4]{\beta}\sqrt{2n\pi}} + O\left(n^{-1}\right) \tag{53}$$

and $O_x\left(n^{\frac{-3}{4}}\right)$ means that $\left\|O_x\left(n^{\frac{-3}{4}}\right)\right\|_{L^2(0,1)} = O\left(n^{\frac{-3}{4}}\right)$.

2. The second family $\{z_{2n}^{+} = (u_{2n}^{+}, \lambda u_{2n}^{+}, v_{2n}^{+}), n \in \mathbb{N}\}$, where z_{2n}^{+} is the eigenfunction of $\mathcal{A}$ corresponding to the eigenvalue λ_{2n}^{+}, has the following asymptotic expression:

$$\left(\partial_x^2 u_{2n}^{+}, \lambda u_{2n}^{+}, v_{2n}^{+}\right) = \left(0,\ \sin\left[\left(n\pi - \frac{\pi}{2}\right)(1-x)\right], 0\right) + O_x\left(n^{-1}\right). \tag{54}$$

3. The third family $\{z_{2n}^{-} = (u_{2n}^{-}, \lambda u_{2n}^{-}, v_{2n}^{-}), n \in \mathbb{N}\}$, where z_{2n}^{-} is the eigenfunction of $\mathcal{A}$ corresponding to the eigenvalue λ_{2n}^{-}, has the following asymptotic expression:

$$\left(\partial_x^2 u_{2n}^{-}, \lambda u_{2n}^{-}, v_{2n}^{-}\right) = \left(\sin\left[\left(n\pi - \frac{\pi}{2}\right)(1-x)\right], 0, 0\right) + O\left(n^{-1}\right). \tag{55}$$

The proof is limited to the first result declared in Theorem 1.3.

Proof: We look for z_{1n} associated with λ_{1n}. From the expression ρ_{1n} given by (45) and a_{1n} given by (46) we have

$$\begin{cases} e^{-a_{1n}y} = e^{\frac{(-1)^{\frac{-11}{8}}\sqrt{\pi x y}}{\sqrt[4]{\beta}} + O\left(n^{-1}\right)}, \quad e^{ia_{1n}y} = e^{\frac{(-1)^{\frac{-7}{8}}\sqrt{\pi x y}}{\sqrt[4]{\beta}} + O\left(n^{-1}\right)}, \\ e^{\pm\sqrt{i}\rho_{1n}(1-x)} = e^{\pm in\pi(1-x) + O\left(n^{-1}\right)}, \end{cases} \tag{56}$$

and the following estimations:

$$\|e^{-a_{1n}y}\| = O\left(n^{\frac{-1}{4}}\right),\ \|e^{ia_{1n}y}\| = O\left(n^{\frac{-1}{4}}\right);$$
$$\left\|e^{\pm\sqrt{i}\rho_{1n}(1-x)}\right\| = O(1),$$

where $y = x$ or $2 - x \in [0, 1]$. According to the matrix $\tilde{X}$ given by (36), for ρ with (45) and a_{1n} given by (46), we obtain

$$u_1 = \frac{1}{e^{2a}e^{\sqrt{i}\rho}} \begin{vmatrix} 1 + e^{2a} & i + ie^{2ia} & 0 \\ (1 - e^{2a})\alpha\rho^2 & (1 - e^{2ia})\alpha\rho^2 & -1 + e^{2\sqrt{i}\rho} \\ e^{ax} - e^{a(2-x)} & e^{iax} - e^{ia(2-x)} & 0 \end{vmatrix}$$
$$= \frac{e^{-\sqrt{i}\rho} - e^{\sqrt{i}}\rho}{\rho^2} \begin{vmatrix} 1 + e^{-2a} & i + ie^{2ia} \\ e^{-a(2-x)} - e^{ax} & e^{iax} - e^{ia(2-x)} \end{vmatrix}.$$

By using estimates (39), we can write

$$u_1 = \frac{e^{-\sqrt{i}\rho} - e^{\sqrt{i}}\rho}{\rho^2} \begin{vmatrix} 1 & i \\ e^{-a(2-x)} - e^{ax} & e^{iax} - e^{ia(2-x)} \end{vmatrix} + O\left(e^{-\gamma_1\sqrt{|\rho|}}\right)$$
$$= \frac{1}{\rho^2}\left(e^{-\sqrt{i}\rho} - e^{\sqrt{i}\rho}\right)\left[\left(e^{iax} - e^{ia(2-x)}\right) - i\left(e^{-a(2-x)} - e^{-ax}\right)\right] + O\left(e^{-\gamma_1\sqrt{|\rho|}}\right).$$

By the expression ρ_{1n} given by (45), we can obtain

$$e^{-\sqrt{i}\rho} - e^{\sqrt{i}\rho} = -2i\sin n\pi + O\left(n^{\frac{-1}{2}}\right) = O\left(n^{\frac{-1}{2}}\right).$$

This together with estimates 77 gives, after a direct computation, that

$$\partial_x^2 u_1 = \frac{a^2}{\rho^2}\left(e^{-\sqrt{i\rho}} - e^{\sqrt{i\rho}}\right)\left[\left(e^{ia(2-x)} - e^{iax}\right) - i\left(e^{-a(2-x)} - e^{-ax}\right)\right] + O\left(e^{-\gamma_1\sqrt{n}}\right) = O_x\left(n^{\frac{-7}{4}}\right),$$

and

$$\lambda u_1 = \left(e^{-\sqrt{i\rho}} - e^{\sqrt{i\rho}}\right)\left[\left(e^{iax} - e^{ia(2-x)}\right) - i\left(e^{-a(2-x)} - e^{-ax}\right)\right] + O\left(e^{-\gamma_1\sqrt{n}}\right) = O_x\left(n^{\frac{-3}{4}}\right).$$

Here,

$O_x\left(n^{\frac{-3}{4}}\right)$ means that $\left\|O_x\left(n^{\frac{-3}{4}}\right)\right\|_{L^2(0,1)} = O\left(n^{\frac{-3}{4}}\right)$ because$\|e^{-ax}\| = \|e^{iax}\| = O\left(n^{\frac{-1}{4}}\right)$.

Similarly, by using estimates (39) and (55), we have

$$\begin{aligned} v_1 &= \frac{1}{e^{2a}e^{\sqrt{i\rho}}\rho^2}\begin{vmatrix} 1+e^{2a} & i+ie^{2ia} & 0 \\ (1-e^{2a})\alpha\rho^2 & \left(1-e^{2ia}\right)\alpha\rho^2 & -1+e^{2\sqrt{i\rho}} \\ 0 & 0 & e^{\sqrt{i\rho}x} - e^{\sqrt{i\rho}(2-x)} \end{vmatrix} \\ &= -2\alpha(1+i)\sin\left[a_n(1-x)\right] + O\left(e^{-\gamma_1\sqrt{n}}\right), \end{aligned}$$

where a_n is given by (52). Let

$$z_{1n} = \frac{-1}{2\alpha(1+i)} z_1,$$

so, we obtain

$$\left(\partial_x^2 u_{1n},\ \lambda u_{1n},\ v_{1n}\right) = \left(0,\ 0,\ \sin\left[a_n(1-x)\right]\right) + O\left(n^{\frac{-3}{4}}\right).$$

The second and third results of Theorem 3 are obtained by the same procedure as before.

Corollary

$$\sigma_r \neq \varnothing. \tag{57}$$

4. Riesz basis property

Lemma.(see [23])

Let $\lambda_n \in \mathbb{C}$, $n = 1, 2, \cdots$, be a sequence that satisfies $\sup_n |\ (\lambda_n)| \leq M$, where M is a positive constant. Then the sine system $\{\sin \lambda_n x, n \geq 1\}$ is a Riesz basis for $L^2(0, 1)$ provided that the sequence λ_n satisfies one of the following conditions:

$$\sup_n |\Re(\lambda_n) - n\pi| < \frac{\pi}{4};$$

$$\sup_n |\Re(\lambda_n) - n\pi + \frac{\pi}{2}| < \frac{\pi}{4}.$$

Lemma.(see [24])

Let $\mathcal{A}$ be a densely defined closed linear operator in a Hilbert space $\mathcal{H}$ with isolated eigenvalues $\{\lambda_i\}_{i=1}^{\infty}$. Let $\{\phi_i\}_{i=1}^{i=\infty}$ be a Riesz basis for $\mathcal{H}$. Suppose that there is an integer $N \geq 1$ and a sequence of generalized eigenvectors $\{\psi_i\}_{i=N}^{\infty}$ of $\mathcal{A}$ such that

$$\sum_{i=N}^{\infty} \|\psi_i - \phi_i\|^2 < \infty.$$

Then, there exists M a number of generalized eigenvectors $\{\psi_{i_0}\}_{i=1}^{M}$ of $\mathcal{A}$ such that

$$\{\psi_{i_0}\}_{i=1}^{M} \cup \{\psi_i\}_{i=M+1}^{\infty}$$

forms a Riesz basis for $\mathcal{H}$.

Theorem 1.4: The generalized eigenfunctions of $\mathcal{A}$ forms a Riesz basis for $\mathcal{H}$. As a result, all eigenvalues with large modules must be algebraically simple and, hence, the spectrum-determined growth condition holds for

$$e^{\mathcal{A}t} : \Phi(\mathcal{A}) = S(\mathcal{A})$$

where

$$\Phi(\mathcal{A}) = \inf\left\{\Phi | \text{there exists an} \, M \text{ such that } \|e^{\mathcal{A}t}\| \leq M e^{\Phi t}\right\},$$

and

$$S(\mathcal{A}) = \sup\{\Re(\lambda) | \lambda \in \sigma(\mathcal{A})\}.$$

Proof: By the bounded invertible mapping:

$$\mathbb{T}(u, w, v) = \left(\partial_x^2 u, w, v\right),$$

the space $\mathcal{H}$ is mapped onto

$$L^2(0, 1) \times L^2(0, 1) \times L^2(0, 1).$$

The value of a_n given by (52) satisfies

$$\sup_n |\, (a_n)| = \sup \left| \frac{\sin \frac{\pi}{8} \alpha^2}{\sqrt{2n\pi \sqrt[4]{\beta}}} \right|$$

is bounded and its real part satisfies

$$\sup_n |\Re(a_n) - n\pi| = \sup_n \left| \frac{\cos \frac{\pi}{8} \alpha^2}{\sqrt{2n\pi \sqrt[4]{\beta}}} \right| \leq \frac{\pi}{4}.$$

Then, it follows that the sequence

$$\{\sin [a_n(1-x)], n = 1, 2, \cdots\},$$

forms a Riesz basis for $L^2(0, 1)$. Similarly, the sequences

$$\left\{ \sin\left[\left(n\pi - \frac{\pi}{2}\right)(1-x)\right], n = 1, 2, \cdots\right\},$$

form a Riesz basis for $L^2(0, 1)$.
Let

$$\Psi_{1n} = (\sin[a_n(1-x)], 0, 0), \Psi_{2n}^{+} = \left(0, \sin\left[\left(n\pi - \frac{\pi}{2}\right)(1-x)\right], 0\right)$$

and

$$\Psi_{2n}^{-} = \left(0, 0, \sin\left[\left(n\pi - \frac{\pi}{2}\right)(1-x)\right]\right).$$

Then, the sequences

$$\{\Psi_{1n}\}_{n\geq 1} \cup \{\Psi_{2n}^{+}\}_{n\geq 1} \cup \{\Psi_{2n}^{-}\}_{n\geq 1}$$

forms a Riesz basis for the following space

$$L^2(0, 1) \times L^2(0, 1) \times L^2(0, 1).$$

Therefore, by the expression of z_{1n}, z_{2n}^{+}, and z_{2n}^{-} given by (51), (53), and (54), respectively, this implies that there exists $N > 0$ such that

$$\sum_{n\geq N}^{\infty}\left[\|\mathbb{T}z_{1n} - \Psi_{1n}\|^2 + \left\|\mathbb{T}z_{2n}^{+} - \Psi_{2n}^{+}\right\|^2 + \left\|\mathbb{T}z_{2n}^{-} - \Psi_{2n}^{-}\right\|^2\right] \leq \sum_{n\geq N}^{\infty} O\left(n^{-2}\right) < \infty.$$

This shows that there is a sequence of generalized eigenfunctions of $\mathcal{A}$, which forms a Riesz basis for $\mathcal{H}$, and all eigenvalues with large modulus must be algebraically simple.

5. Exponential stability

Theorem 1.5: The C_0−semigroup $S(t)$ generated by the operator $\mathcal{A}$ is exponentially stable, that is,

$$\left\|e^{\mathcal{A}t}\right\| \leq Me^{\omega t},$$

where M and ω are positive constants[6].

Proof: By the asymptotic distribution of eigenvalues given by Theorem 1.2 and the continuous spectrum given by Eq. (50), in addition to the empty residual spectrum set given by Eq. (56), we conclude that $S(\mathcal{A}) = -\frac{1}{\beta}$. The proof is completed by the spectrum-determined growth condition, which is similar to [24–26].

[6] By recalling the eigenvalues of $\mathcal{A}$ given by 44, we deduce that $\omega \geq -\frac{1}{\beta}$.

6. Conclusion

The main results of this work are similar to those mentioned in [27], the results are summarized as follows:

1. The system operator of the closed-loop system is not of compact resolvent and the spectrum consists of three branches.

2. By means of asymptotic analysis, the asymptotic expressions of eigenfunctions are obtained.

3. By the comparison method in the Riesz basis approach, exponential stability is obtained.

Conflict of interest

The authors declare no conflict of interest.

Author details

Hanni Dridi
Applied Mathematics Laboratory, University of Badji Mokhtar, Annaba, Algeria

*Address all correspondence to: hannidridi@gmail.com

References

[1] Zhang X, Zuazua E. Polynomial decay and control of a 1–d hyperbolic–parabolic coupled system. Journal of Differential Equations. 2004;**204**(2): 380-438

[2] Zhang X, Zuazua E. Asymptotic behavior of a hyperbolic-parabolic coupled system arising in fluid-structure interaction. In: Free Boundary Problems. Basel: Birkhäuser; 2006. pp. 445-455

[3] Zhang Q, Wang J-M, Guo B-Z. Stabilization of the Euler–Bernoulli equation via boundary connection with heat equation. Mathematics of Control, Signals, and Systems. 2014;**26**(1):77-118

[4] Wang J-M, Krstic M. Stability of an interconnected system of euler–bernoulli beam and heat equation with boundary coupling. ESAIM. 2015;**21**(4): 1029-1052

[5] Wang J-M, Ren B, Krstic M. Stabilization and Gevrey regularity of a Schrödinger equation in boundary feedback with a heat equation. IEEE Transactions on Automatic Control. 2011;**57**(1):179-185

[6] Guo BZ. Further results for a one-dimensional linear thermoelastic equation with Dirichlet-Dirichlet boundary conditions. The ANZIAM Journal. 2002;**43**(3):449-462

[7] Dridi H, Djebabla A. On the stabilization of linear porous elastic materials by microtemperature effect and porous damping. Annali Dell'Universita'di ferrara. 2020;**66**(1): 13-25

[8] Dridi H. Timoshenko system with fractional operator in the memory and spatial fractional thermal effect. Rendiconti del Circolo Matematico di Palermo Series. 2021;**70**(1):593-621

[9] Hanni D, Feng B, Zennir K. Stability of Timoshenko system coupled with thermal law of Gurtin-Pipkin affecting on shear force. Applicable Analysis. 2021;**2021**:1-22

[10] Dridi H, Zennir K. Well-posedness and energy decay for some thermoelastic systems of Timoshenko type with Kelvin–Voigt damping. SeMA Journal. 2021;**78**(3):385-400

[11] Dridi H. Decay rate estimates for a new class of multidimensional nonlinear Bresse systems with time-dependent dissipations. Ricerche di Matematica. 2021;**2021**:1-33

[12] Dridi H, Saci M, Djebabla A. General decay of Bresse system by modified thermoelasticity of type III. Annali Dell'universita'di ferrara. 2022;**68**(1): 203-222

[13] Hanni D, Djebabla A, Tatar N. Well-posedness and exponential stability for the von Karman systems with second sound. Eurasian Journal of Mathematical and Computer Applications. 2019;**7**(4):52-65

[14] West H. The Euler-Bernoulli beam equation with boundary energy dissipation. Operator methods for optimal control problems. 1987

[15] Huang F. Characteristic conditions for exponential stability of linear dynamical systems in Hilbert spaces. Annals of Differential Equations. 1985;**1**: 43-56

[16] Chen G. Energy decay estimates and exact boundary value controllability for the wave equation in a bounded domain. 1979

[17] Lagnese J. Decay of solutions of wave equations in a bounded region with

boundary dissipation. Journal of Differential Equations. 1983;**50**(2): 163-182

[18] Wazwaz A-M. A study on linear and nonlinear Schrodinger equations by the variational iteration method. Chaos, Solitons & Fractals. 2008;**37**(4): 1136-1142

[19] He J-H. Some asymptotic methods for strongly nonlinear equations. International Journal of Modern Physics B. 2006;**20**(10):1141-1199

[20] He J-H. Variational iteration method–a kind of non-linear analytical technique: Some examples. International Journal of Non-Linear Mechanics. 1999; **34**(4):699-708

[21] Hanni D, Khaled Z. New Class of Kirchhoff Type Equations with Kelvin-Voigt Damping and General Nonlinearity: Local Ex-istence and Blow-up in Solutions. Journal of Partial Differential Equations. 2021;**34**(4): 313-347

[22] Pazy A. Semigroups of Linear Operators and Applications to Partial Differential Equations. Vol. 44. Springer Science Business Media. 2012

[23] Bilalov B. Bases of exponentials, cosines, and sines formed by eigenfunctions of differential operators. Differential Equations. 2003;**39**(5):652-657

[24] Guo B-Z, Guo-Dong Z. On Spectrum and Riesz basis property for one-dimensional wave equation with Boltzmann damping. ESAIM: Control Optimisation and Calculus of Variations. 2012;**18**(3):889-913

[25] Guo B-Z, Chan KY. Riesz basis generation, eigenvalues distribution, and exponential stability for a Euler-Bernoulli beam with joint feedback control. Revista Matemática Complutense. 2001;**14**(1):205-229

[26] Guo B-Z. Riesz basis approach to the stabilization of a flexible beam with a tip mass. SIAM Journal on Control and Optimization. 2001;**39**(6):1736-1747

[27] Guo B-Z, Ren H-J. Stabilization and regularity transmission of a Schrödinger equation through boundary connections with a Kelvin-Voigt damped beam equation. ZAMM-Journal of Applied Mathematics and Mechanics/Zeitschrift für Angewandte Mathematik und Mechanik. 2020;**100**(2):e201900013

Chapter 6

Generalized Bessel Operator and Reproducing Kernel Theory

Fethi Soltani

Abstract

In 1961, Bargmann introduced the classical Fock space $F(\mathbb{C})$ and in 1984, Cholewinsky introduced the generalized Fock space $F_{2,\nu}(\mathbb{C})$. These two spaces are the aim of many works, and have many applications in mathematics, in physics and in quantum mechanics. In this work, we introduce and study the Fock space $F_{3,\nu}(\mathbb{C})$ associated to the generalized Bessel operator $L_{3,\nu}$. The space $F_{3,\nu}(\mathbb{C})$ is a reproducing kernel Hilbert space (RKHS). This is the reason for defining the orthogonal projection operator, the Toeplitz operators and the Hankel operators associated to this space. Furthermore, we give an application of the theory of extremal function and reproducing kernel of Hilbert space, to establish the extremal function associated to a bounded linear operator $T : F_{3,\nu}(\mathbb{C}) \to H$, where H be a Hilbert space. Finally, we come up with some results regarding the extremal functions, when T is a difference operator and an integral operator, respectively. Finally, we remark that it is now natural to raise the problem of studying the Bessel-type Segal-Bargmann transform associated to the space $F_{3,\nu}(\mathbb{C})$. This problem is difficult and will be an open topic. This topic requires more details for the harmonic analysis associated to the operator $L_{3,\nu}$. We have the idea to continue this research in a future paper.

Keywords: Bessel-type Fock space, Heisenberg-type uncertainty principle, Toeplitz operators, Hankel operators, Tikhonov regularization problem, extremal function

1. Introduction

In [1], Bargmann has studied the Fock space $F(\mathbb{C})$, is a Hilbert space consisting of entire functions on $\mathbb{C}$, square integrable with respect to the measure

$$\mathrm{d}m(z) := \frac{1}{\pi} e^{-|z|^2} \mathrm{d}x\mathrm{d}y, \quad z = x + iy.$$

This space is equipped with the inner product

$$\langle f, g \rangle_{F(\mathbb{C})} := \int_{\mathbb{C}} f(z)\overline{g(z)}\mathrm{d}m(z),$$

and has the reproducing kernel $k(z, w) = e^{\bar{z}w}$. In [2], Cholewinsky has constructed a generalized Fock space $F_{2,\nu}(\mathbb{C})$ consisting of even entire functions on $\mathbb{C}$, square integrable with respect to the measure

$$dm_{2,\nu}(z) := \frac{|z|^{2\nu+1} K_{\nu-1/2}\left(|z|^2\right)}{\pi 2^{\nu-1/2}\Gamma(\nu+1/2)} dz,$$

where K_ν, $\nu > 0$, is the modified Bessel function of the second kind and index ν, called also the Macdonald function [3]. The generalized Fock space $F_{2,\nu}(\mathbb{C})$ is associated to the Bessel operator

$$L_{2,\nu} := \frac{d^2}{dz^2} + \frac{2\nu}{z}\frac{d}{dz},$$

and has the reproducing kernel

$$k_{2,\nu}(z,w) = I_{2,\nu}(\bar{z}w) = \sum_{n=0}^{\infty} \frac{(\bar{z}w)^{2n}}{\alpha_n(2,\nu)},$$

where

$$\alpha_n(2,\nu) = 2^{2n} n! \frac{\Gamma(n+\nu+1/2)}{\Gamma(\nu+1/2)}.$$

The study of several generalizations of the Fock spaces has a long and rich history in many different settings [4–8]. In this work, we will try to generalize Bessel-type Fock space, to give some properties concerning Toeplitz operators and Hankel operators of this space; and to establish Heisenberg-type uncertainty principle for this generalized Fock space. The generalized Bessel operator (or hyper-Bessel operator [9]) is the third-order singular differential operator given by

$$L_{3,\nu} := \frac{d^3}{dz^3} + \frac{3\nu}{z}\frac{d^2}{dz^2} - \frac{3\nu}{z^2}\frac{d}{dz},$$

where ν is a nonnegative real number. When $\nu = 0$ this operator becomes the third derivative operator for which some analysis was studied by Widder [10] and for some special value of ν the operator $L_{3,\nu}$ appeared as a radial part of the generalized Airy equation of a nonlinear diffusion type partial differential equation in $\mathbb{R}^d$. Recently, in a nice and long paper, Cholewinski and Reneke [11] studied and extended, for the operator $L_{3,\nu}$, the well known theory related to some singular differential operator of second order for which the literature is extensive. Next, Fitouhi et al. [12, 13] established a harmonic analysis related to this operator (for examples the eigenfunctions, the generalized translation, the Fourier-Airy transform, the heat equation, the heat polynomials, the transmutation operators, ...). Recently the Airy operator has gained considerable interest in various field of mathematics [9, 14] and in certain parts of quantum mechanics [15]. The results of this work will be useful when discussing the Fock space associated to this operator. This space is the background of some applications in this contribution. Especially, we give an application of the theory of extremal functions and reproducing kernels of Hilbert spaces, to examine the extremal function for the Tikhonov regularization problem associated to a bounded linear operator $T : F_{3,\nu}(\mathbb{C}) \to H$, where H be a Hilbert space. We come up with some results regarding the extremal functions associated to a difference operator D and to an integral operator P.

The examination of the extremal functions is studied in several directions, in the Fourier analysis [16, 17], in the Sturm-Liouville hypergroups [18], and in the Fourier-Dunkl analysis [19, 20] ...

The contents of the paper are as follows. In Section 2, we study the Toeplitz operators and the Hankel operators on the Bessel-type Fock space $F_{3,\nu}(\mathbb{C})$, and we establish Heisenberg-type uncertainty principle for this space. In Section 3, we give an application of the theory of reproducing kernels to the Tikhonov regularization problem for a difference operator and for an integral operator, respectively. In the last section, we summarize the obtained results and describe the future work.

2. Bessel-type Fock space

In this section we introduce the Toeplitz and the Hankel operators on the Bessel-type Fock space $F_{3,\nu}(\mathbb{C})$. And we establish an uncertainty inequality of Heisenberg-type on the space $F_{3,\nu}(\mathbb{C})$.

2.1 Toeplitz and Hankel operators on $F_{3,\nu}(\mathbb{C})$

Let $z\in\mathbb{C}$ and $\omega_k = e^{2i\pi\frac{k-1}{3}}$, $k = 1, 2, 3$. A function $u(z)$ is called 3-even if $u(\omega_k z) = u(z)$.

For $\lambda\in\mathbb{C}$, the initial problem

$$L_{3,\nu}u(z) = \lambda^3 u(z), \quad u(0) = 1, u^{(k)}(0) = 0, k = 1, 2$$

admits a unique analytic solution on $\mathbb{C}$ (see [11]), which will be denoted by $I_{3,\nu}(\lambda z)$ and expanded in a power series as

$$I_{3,\nu}(\lambda z) = \sum_{n=0}^{\infty} \frac{(\lambda z)^{3n}}{\alpha_n(3,\nu)}, \tag{1}$$

where

$$\alpha_n(3,\nu) = 3^{3n} n! \frac{\Gamma(n+1/3)\Gamma(n+\nu+2/3)}{\Gamma(1/3)\Gamma(\nu+2/3)}.$$

The function $I_{3,\nu}(\lambda z)$ is 3-even and defined as the hypergeometric function [11],

$$I_{3,\nu}(\lambda z) = {}_0F_2\left[\frac{1}{3}, \nu + \frac{2}{3} \middle| \left(\frac{\lambda z}{3}\right)^3\right].$$

In particular, $|I_{3,\nu}(\lambda z)| \le e^{|\lambda||z|}$ and $I_{3,0}(\lambda z) = \cos_3(-\lambda z) = \sum_{n=0}^{\infty} \frac{(\lambda z)^{3n}}{(3n)!}$.

In the following we denote by

- O_ν the function (see [11], p. 12) defined for $t \ge 0$ by

$$O_\nu(t) := \frac{2t^{1/3}}{3^{(3\nu+5)/2}\Gamma(1/3)\Gamma(\nu+2/3)} \int_0^{\infty} x^{(\nu-1)/2} e^{-t/x} K_{\nu-1/3}\left(\sqrt{\frac{4x}{27}}\right) dx,$$

where K_ν is the Macdonald function.

- $dm_{3,\nu}(z)$, $\nu > 0$, the measure defined on $\mathbb{C}$ by

$$dm_{3,\nu}(z) := \frac{3}{\pi}|z|^{-2}O_\nu\left(|z|^6\right)dxdy, \quad z = x + iy.$$

This weighted measure is use d by Cholewinski-Reneke in their interesting paper [11] for computing the ν-Airy heat function.

- $L^2_\nu(\mathbb{C})$, the space of measurable functions f on $\mathbb{C}$ satisfying

$$\|f\|^2_{L^2_\nu(\mathbb{C})} := \int_{\mathbb{C}} |f(z)|^2 dm_{3,\nu}(z) < \infty.$$

- $H_{3,*}(\mathbb{C})$, the space of 3-even entire functions on $\mathbb{C}$.

Let $\nu > 0$. We define the Bessel-type Fock space $F_{3,\nu}(\mathbb{C})$, to be the pre-Hilbert space of functions in $H_{3,*}(\mathbb{C}) \cap L^2_\nu(\mathbb{C})$, equipped with the inner product

$$\langle f, g\rangle_{F_{3,\nu}(\mathbb{C})} := \int_{\mathbb{C}} f(z)\overline{g(\overline{z})}dm_{3,\nu}(z),$$

and the norm $\|f\|_{F_{3,\nu}(\mathbb{C})} := \|f\|_{L^2_\nu(\mathbb{C})}$.
If $f, g \in F_{3,\nu}(\mathbb{C})$ with $f(z) = \sum_{n=0}^{\infty} a_n z^{3n}$ and $g(z) = \sum_{n=0}^{\infty} b_n z^{3n}$, then

$$\langle f, g\rangle_{F_{3,\nu}(\mathbb{C})} = \sum_{n=0}^{\infty} a_n \overline{b_n} \alpha_n(3, \nu), \tag{2}$$

where $\alpha_n(3, \nu)$ are the constants given by (1).
If $f \in F_{3,\nu}(\mathbb{C})$, then $|f(z)| \leq e^{|z|^2/2}\|f\|_{F_{3,\nu}(\mathbb{C})}$, $z \in \mathbb{C}$. The map $f \to f(z)$, $z \in \mathbb{C}$, is a continuous linear functional on $F_{3,\nu}(\mathbb{C})$. Thus from Riesz theorem [21], the space $F_{3,\nu}(\mathbb{C})$ has a reproducing kernel. The function $k_{3,\nu}$ given for, by.

$$k_{3,\nu}(z, w) = I_{3,\nu}(\overline{z}w), \tag{3}$$

is a reproducing kernel for the Bessel-type Fock space $F_{3,\nu}(\mathbb{C})$, that is $k_{3,\nu}(z, .) \in F_{3,\nu}(\mathbb{C})$, and for all $f \in F_{3,\nu}(\mathbb{C})$, we have $\langle f, k_{3,\nu}(z, .)\rangle_{F_{3,\nu}(\mathbb{C})} = f(z)$.
The space $F_{3,\nu}(\mathbb{C})$ equipped with the inner product $\langle ., .\rangle_{F_{3,\nu}(\mathbb{C})}$ is a reproducing kernel Hilbert space (RKHS); and the set $\left\{\frac{z^{3n}}{\sqrt{\alpha_n(3,\nu)}}\right\}_{n\in\mathbb{N}}$ forms a Hilbert's basis for the space $F_{3,\nu}(\mathbb{C})$.
In the next part of this section we study the Toeplitz operators and the Hankel operators on the Bessel-type Fock space $F_{3,\nu}(\mathbb{C})$. These operators generalize the classical operators [5]. We consider the orthogonal projection operator $P_\nu : L^2_\nu(\mathbb{C}) \to F_{3,\nu}(\mathbb{C})$ defined for $z \in \mathbb{C}$, by

$$P_\nu f(z) := \langle f, k_{3,\nu}(z, .)\rangle_{L^2_\nu(\mathbb{C})},$$

where $k_{3,\nu}$ is the reproducing kernel given by (3). Then we have

$$P_\nu \circ P_\nu = P_\nu, \quad P_\nu^* = P_\nu, \quad \|P_\nu\| = 1, \quad \|I - P_\nu\| \leq 1.$$

Let $\phi \in L^\infty(\mathbb{C})$. The multiplication operators $M_\phi : L^2_\nu(\mathbb{C}) \to L^2_\nu(\mathbb{C})$ are the operators defined for $z \in \mathbb{C}$, by

$$M_\phi f(z) := \phi(z) f(z).$$

The Bessel-type Toeplitz operators $T_\phi : F_{3,\nu}(\mathbb{C}) \to F_{3,\nu}(\mathbb{C})$ are the operators defined for $z \in \mathbb{C}$, by

$$T_\phi f(z) := P_\nu M_\phi f(z).$$

Let $\phi \in L^\infty(\mathbb{C})$. Then we have

$$\|T_\phi\| \leq \|\phi\|_\infty, \quad T_\phi^* = T_{\overline{\phi}}.$$

However, if $\phi \in L^\infty(\mathbb{C})$ has compact support, then T_ϕ is a compact operator.

Let $\phi \in L^\infty(\mathbb{C})$. The Bessel-type Hankel operators $H_\phi : F_{3,\nu}(\mathbb{C}) \to L^2_\nu(\mathbb{C})$ are the operators defined for $z \in \mathbb{C}$, by

$$H_\phi f(z) := (I - P_\nu) M_\phi f(z).$$

Let $\phi, \varphi \in L^\infty(\mathbb{C})$. Then we have

$$\|H_\phi\| \leq \|\phi\|_\infty, \quad H_\phi^* = P_\nu M_{\overline{\phi}}(I - P_\nu), \quad T_{\phi\varphi} - T_\phi T_\varphi = H_{\overline{\phi}}^* H_\varphi.$$

2.2 Heisenberg-type uncertainty principle for $F_{3,\nu}(\mathbb{C})$

Let $U_{3,\nu}(\mathbb{C})$ be the prehilbertian space of 3-even entire functions, equipped with the inner product

$$\langle f, g \rangle_{U_{3,\nu}(\mathbb{C})} := \int_{\mathbb{C}} f(z) \overline{g(z)} |z|^6 dm_{3,\nu}(z).$$

If $f, g \in U_{3,\nu}(\mathbb{C})$ with $f(z) = \sum_{n=0}^\infty a_n z^{3n}$ and $g(z) = \sum_{n=0}^\infty b_n z^{3n}$, then

$$\langle f, g \rangle_{U_{3,\nu}(\mathbb{C})} = \sum_{n=0}^\infty a_n \overline{b_n} \alpha_{n+1}(3,\nu), \quad \|f\|^2_{U_{3,\nu}(\mathbb{C})} = \sum_{n=0}^\infty |a_n|^2 \alpha_{n+1}(3,\nu).$$

The space $U_{3,\nu}(\mathbb{C})$ is a Hilbert space with Hilbert's basis $\left\{ \frac{z^{3n}}{\sqrt{\alpha_{n+1}(3,\nu)}} \right\}_{n \in \mathbb{N}}$ and reproducing kernel

$$J_{3,\nu}(z,w) = \sum_{n=0}^\infty \frac{(\overline{z}w)^{3n}}{\alpha_{n+1}(3,\nu)} = \frac{1}{(\overline{z}w)^3} (I_{3,\nu}(\overline{z}w) - 1).$$

Using the fact that

$$\alpha_{n+1}(3,\nu) = 3(n+1)(3n+1)(3n+3\nu+2)\alpha_n(3,\nu) \geq \alpha_n(3,\nu), \quad n \in \mathbb{N}, \tag{4}$$

then the space $U_{3,\nu}(\mathbb{C})$ is a subspace of the Bessel-type Fock space $F_{3,\nu}(\mathbb{C})$.
Let M be the multiplication operator defined by

$$Mf(z) := z^3 f(z).$$

Lemma 1. For $f \in U_{3,\nu}(\mathbb{C})$ then $L_{3,\nu}f$ and Mf belong to $F_{3,\nu}(\mathbb{C})$. And for $f, g \in U_{3,\nu}(\mathbb{C})$ one has

$$\langle L_{3,\nu}f, g\rangle_{F_{3,\nu}(\mathbb{C})} = \langle f, Mg\rangle_{F_{3,\nu}(\mathbb{C})}.$$

Proof. Let $f \in U_{3,\nu}(\mathbb{C})$ with $f(z) = \sum_{n=0}^{\infty} a_n z^{3n}$. By straightforward calculation we obtain

$$L_{3,\nu}\left(z^{3n}\right) = 3n(3n-2)(3n+3\nu-1)z^{3n-3}, \quad n \in \mathbb{N}^*.$$

Thus

$$L_{3,\nu}f(z) = \sum_{n=0}^{\infty} 3(n+1)(3n+1)(3n+3\nu+2)a_{n+1}z^{3n}, \tag{5}$$

and by (4) we deduce that

$$\|L_{3,\nu}f\|^2_{F_\nu(\mathbb{C})} = \sum_{n=1}^{\infty} 3n(3n-2)(3n+3\nu-1)|a_n|^2 \alpha_{3n}(\nu) \leq \|f\|^2_{U_{3,\nu}(\mathbb{C})}.$$

On the other hand for Mf one has

$$Mf(z) = \sum_{n=1}^{\infty} a_{n-1} z^{3n},$$

and

$$\|Mf\|^2_{F_{3,\nu}(\mathbb{C})} = \sum_{n=1}^{\infty} |a_{n-1}|^2 \alpha_n(3,\nu) = \|f\|^2_{U_{3,\nu}(\mathbb{C})}.$$

Therefore $L_{3,\nu}f$ and Mf belong to $F_{3,\nu}(\mathbb{C})$.

Let $f, g \in U_{3,\nu}(\mathbb{C})$ with $f(z) = \sum_{n=0}^{\infty} a_n z^{3n}$ and $g(z) = \sum_{n=0}^{\infty} b_n z^{3n}$. From relations (4) and (5) we have

$$L_{3,\nu}f(z) = \sum_{n=0}^{\infty} a_{n+1} \frac{\alpha_{n+1}(3,\nu)}{\alpha_n(3,\nu)} z^{3n}.$$

Therefore and according to (2) we obtain

$$\langle L_{3,\nu}f, g\rangle_{F_{3,\nu}(\mathbb{C})} = \sum_{n=0}^{\infty} a_{n+1}\overline{b_n}\alpha_{n+1}(3,\nu) = \sum_{n=1}^{\infty} a_n \overline{b_{n-1}} \alpha_{3n}(\nu) = \langle f, Mg\rangle_{F_{3,\nu}(\mathbb{C})}.$$

This completes the proof of the lemma. □

Let $V_{3,\nu}(\mathbb{C})$ be the prehilbertian space of 3-even entire functions, equipped with the inner product

$$\langle f,g\rangle_{V_{3,\nu}(\mathbb{C})} := \int_{\mathbb{C}} f(z)\overline{g(z)}|z|^{12}dm_{3,\nu}(z).$$

If $f,g \in V_{3,\nu}(\mathbb{C})$ with $f(z) = \sum_{n=0}^{\infty} a_n z^{3n}$ and $g(z) = \sum_{n=0}^{\infty} b_n z^{3n}$, then

$$\langle f,g\rangle_{V_{3,\nu}(\mathbb{C})} = \sum_{n=0}^{\infty} a_n\overline{b_n}\alpha_{n+2}(3,\nu), \quad \|f\|^2_{V_{3,\nu}(\mathbb{C})} = \sum_{n=0}^{\infty} |a_n|^2\alpha_{n+2}(3,\nu).$$

The space $V_{3,\nu}(\mathbb{C})$ is a Hilbert space with Hilbert's basis $\left\{\frac{z^{3n}}{\sqrt{\alpha_{n+2}(3,\nu)}}\right\}_{n\in\mathbb{N}}$ and reproducing kernel

$$K_{3,\nu}(z,w) = \sum_{n=0}^{\infty} \frac{(\overline{z}w)^{3n}}{\alpha_{n+2}(3,\nu)} = \frac{1}{(\overline{z}w)^6}\left(I_{3,\nu}(\overline{z}w) - \frac{(\overline{z}w)^3}{9\nu+6} - 1\right).$$

The space $V_{3,\nu}(\mathbb{C})$ is a subspace of the space $U_{3,\nu}(\mathbb{C})$.
Let $[L_{3,\nu}, M]$ the commutator operator defined by

$$[L_{3,\nu}, M] := L_{3,\nu}M - ML_{3,\nu}.$$

We easily have
Lemma 2. For $f \in V_{3,\nu}(\mathbb{C})$ we have $[L_{3,\nu}, M]f \in F_{3,\nu}(\mathbb{C})$ and

$$\|Mf\|^2_{F_{3,\nu}(\mathbb{C})} = \|L_{3,\nu}f\|^2_{F_{3,\nu}(\mathbb{C})} + \langle [L_{3,\nu}, M]f, f\rangle_{F_{3,\nu}(\mathbb{C})}.$$

We will use the following result of functional analysis.
Lemma 3. (See [22, 23]). Let A and B be self-adjoint operators on a Hilbert space H $(A^* = A, B^* = B)$. Then we have

$$\|(A-a)f\|_H\|(B-b)f\|_H \geq \frac{1}{2}|\langle [A,B]f,f\rangle_H|,$$

for all $f \in Dom\,([A,B])$, and all $a,b \in \mathbb{R}$.
We obtain the following Heisenberg-type uncertainty principle.
Theorem 1. Let $f \in V_{3,\nu}(\mathbb{C})$. For all $a,b \in \mathbb{R}$, we have.

$$\|(L_{3,\nu} + M - a)f\|_{F_{3,\nu}(\mathbb{C})}\|(L_{3,\nu} - M + ib)f\|_{F_{3,\nu}(\mathbb{C})} \geq \left|\|Mf\|^2_{F_{3,\nu}(\mathbb{C})} - \|L_{3,\nu}f\|^2_{F_{3,\nu}(\mathbb{C})}\right|. \quad (6)$$

Proof. Let us consider the following two operators on $V_{3,\nu}(\mathbb{C})$ by

$$A = L_{3,\nu} + M, \quad B = i(L_{3,\nu} - M).$$

By Lemmas 1 and 2, the operators A and B satisfies the following properties.

i. For $f,g \in V_{3,\nu}(\mathbb{C})$, we have

$$\langle Af,g\rangle_{F_{3,\nu}(\mathbb{C})} = \langle f,Ag\rangle_{F_{3,\nu}(\mathbb{C})}, \quad \langle Bf,g\rangle_{F_{3,\nu}(\mathbb{C})} = \langle f,Bg\rangle_{F_{3,\nu}(\mathbb{C})}.$$

$$V_{3,\nu}(\mathbb{C}) \subset Dom([A,B]).$$
$$[A,B] = -2i[L_{3,\nu}, M].$$

Thus the inequality (6) follows from Lemmas 2 and 3. □

3. Reproducing kernel theory

Let $T : F_{3,\nu}(\mathbb{C}) \to H$ be a bounded linear operator from $F_{3,\nu}(\mathbb{C})$ into a Hilbert space H. By using the theory of reproducing kernels of Hilbert space and building on the ideas of Saitoh [24–26] we examine the extremal function associated to the operator T on the Bessel-type Fock space $F_{3,\nu}(\mathbb{C})$.

Theorem 2. For any $h \in H$ and for any $\lambda > 0$, the problem

$$\inf_{f \in F_{3,\nu}(\mathbb{C})} \left\{ \lambda \|f\|^2_{F_{3,\nu}(\mathbb{C})} + \|Tf - h\|^2_H \right\} \tag{7}$$

has a unique minimizer given by

$$f^*_{\lambda,T}(h) = (\lambda I + T^*T)^{-1} T^* h. \tag{8}$$

Proof. The problem (7) is solved elementarily by finding the roots of the first derivative $d\Phi$ of the quadratic and strictly convex function $\Phi(f) = \lambda\|f\|^2_{F_{3,\nu}(\mathbb{C})} + \|Tf - h\|^2_H - \|h\|^2_H$. Note that for convex functions the equation $d\Phi(f) = 0$ is a necessary and sufficient condition for the minimum at f. The calculation provides

$$d\Phi(f) = 2\lambda f + 2T^*(Tf - h),$$

and the assertion of the theorem follows at once. □

In this section we examine the extremal functions associated to a difference operator D; and to an integral operator P, respectively.

3.1 The difference operator

Let D be the difference operator defined by

$$Df(z) := \frac{1}{z^3}(f(z) - f(0)).$$

If $f(z) = \sum_{n=0}^{\infty} a_n z^{3n}$, then $Df(z) = \sum_{n=0}^{\infty} a_{n+1} z^{3n}$.

In this subsection, we determine the extremal function $f^*_{\lambda,D}$ associated to the difference operator D on the space $F_{3,\nu}(\mathbb{C})$.

Theorem 3.

i. The operator D maps continuously from $F_{3,\nu}(\mathbb{C})$ into $F_{3,\nu}(\mathbb{C})$, and

$$\|Df\|_{F_{3,\nu}(\mathbb{C})} \le \|f\|_{F_{3,\nu}(\mathbb{C})}.$$

ii. For $f \in F_{3,\nu}(\mathbb{C})$ with $f(z) = \sum_{n=0}^{\infty} a_n z^{3n}$, we have

$$D^* f(z) = \sum_{n=1}^{\infty} \frac{\alpha_{n-1}(3,\nu)}{\alpha_n(3,\nu)} a_{n-1} z^{3n}, \quad D^* Df(z) = \sum_{n=1}^{\infty} \frac{\alpha_{n-1}(3,\nu)}{\alpha_n(3,\nu)} a_n z^{3n}.$$

iii. For any $h \in F_{3,\nu}(\mathbb{C})$ and for any $\lambda > 0$, the problem

$$\inf_{f \in F_{3,\nu}(\mathbb{C})} \left\{ \lambda \|f\|^2_{F_{3,\nu}(\mathbb{C})} + \|Df - h\|^2_{F_{3,\nu}(\mathbb{C})} \right\}$$

has a unique minimizer given by

$$f^*_{\lambda,D}(h)(z) = \langle h, \Psi_z \rangle_{F_{3,\nu}(\mathbb{C})},$$

where

$$\Psi_z(w) = \sum_{n=0}^{\infty} \frac{(\overline{z})^{3n+3} w^{3n}}{\lambda \alpha_{n+1}(3,\nu) + \alpha_n(3,\nu)}, \quad w \in \mathbb{C}.$$

Proof.

i. If $f \in F_{3,\nu}(\mathbb{C})$ with $f(z) = \sum_{n=0}^{\infty} a_n z^{3n}$, then

$$\|Df\|^2_{F_{3,\nu}(\mathbb{C})} = \sum_{n=0}^{\infty} \alpha_n(3,\nu)|a_{n+1}|^2 \le \sum_{n=1}^{\infty} \alpha_n(3,\nu)|a_n|^2 \le \|f\|^2_{F_{3,\nu}(\mathbb{C})}.$$

ii. If $f, g \in F_{3,\nu}(\mathbb{C})$ with $f(z) = \sum_{n=0}^{\infty} a_n z^{3n}$ and $g(z) = \sum_{n=0}^{\infty} b_n z^{3n}$, then

$$\langle Df, g \rangle_{F_{3,\nu}(\mathbb{C})} = \sum_{n=0}^{\infty} \alpha_n(3,\nu) a_{n+1} \overline{b_n} = \sum_{n=1}^{\infty} \alpha_{n-1}(3,\nu) a_n \overline{b_{n-1}} = \langle f, D^* g \rangle_{F_{3,\nu}(\mathbb{C})},$$

where

$$D^* g(z) = \sum_{n=1}^{\infty} \frac{\alpha_{n-1}(3,\nu)}{\alpha_n(3,\nu)} b_{n-1} z^{3n}.$$

And therefore

$$D^* Df(z) = \sum_{n=1}^{\infty} \frac{\alpha_{n-1}(3,\nu)}{\alpha_n(3,\nu)} a_n z^{3n}.$$

iii. We put $h(z) = \sum_{n=0}^{\infty} h_n z^{3n}$ and $f^*_{\lambda,D}(h)(z) = \sum_{n=0}^{\infty} c_n z^{3n}$. From (8) we have $(\lambda I + D^* D) f^*_{\lambda,D}(h)(z) = D^* h(z)$. By (ii) we deduce that

$$c_0 = 0, \quad c_n = \frac{\alpha_{n-1}(3,\nu) h_{n-1}}{\lambda \alpha_n(3,\nu) + \alpha_{n-1}(3,\nu)}, \quad n \in \mathbb{N}^*.$$

Thus

$$f^*_{\lambda,D}(h)(z) = \sum_{n=0}^{\infty} \frac{\alpha_n(3,\nu)h_n z^{3n+3}}{\lambda\alpha_{n+1}(3,\nu) + \alpha_n(3,\nu)} = \langle h, \Psi_z \rangle_{F_{3,\nu}(\mathbb{C})}, \tag{9}$$

where

$$\Psi_z(w) = \sum_{n=0}^{\infty} \frac{(\overline{z})^{3n+3} w^{3n}}{\lambda\alpha_{n+1}(3,\nu) + \alpha_n(3,\nu)}.$$

This completes the proof of the theorem. □
The extremal function $f^*_{\lambda,D}(h)$ possesses the following properties.
Theorem 4. If $\lambda > 0$ and $h \in F_{3,\nu}(\mathbb{C})$, then

$$|f^*_{\lambda,D}(h)(z)| \leq \frac{1}{2\sqrt{\lambda}} \left(I_{3,\nu}\left(|z|^2\right)\right)^{1/2} \|h\|_{F_{3,\nu}(\mathbb{C})},$$

$$\|f^*_{\lambda,D}(h)\|_{F_{3,\nu}(\mathbb{C})} \leq \frac{1}{2\sqrt{\lambda}} \|h\|_{F_{3,\nu}(\mathbb{C})}.$$

Proof. Let $\lambda > 0$ and $h \in F_{3,\nu}(\mathbb{C})$ with $h(z) = \sum_{n=0}^{\infty} h_n z^{3n}$.

i. From (9) we have

$$|f^*_{\lambda,D}(h)(z)| \leq \|\Psi_z\|_{F_{3,\nu}(\mathbb{C})} \|h\|_{F_{3,\nu}(\mathbb{C})}.$$

And by using the fact that $(x+y)^2 \geq 4xy$ we obtain

$$\|\Psi_z\|^2_{F_{3,\nu}(\mathbb{C})} = \sum_{n=0}^{\infty} \frac{\alpha_n(3,\nu)|z|^{6n+6}}{[\lambda\alpha_{n+1}(3,\nu) + \alpha_n(3,\nu)]^2} \leq \frac{1}{4\lambda} \sum_{n=0}^{\infty} \frac{|z|^{6n+6}}{\alpha_{n+1}(3,\nu)} \leq \frac{1}{4\lambda} I_{3,\nu}\left(|z|^2\right).$$

This proves (i).

ii. From (9) we have

$$\|f^*_{\lambda,D}(h)\|^2_{F_{3,\nu}(\mathbb{C})} = \sum_{n=1}^{\infty} \alpha_n(3,\nu) \left[\frac{\alpha_{n-1}(3,\nu)|h_{n-1}|}{\lambda\alpha_n(3,\nu) + \alpha_{n-1}(3,\nu)}\right]^2.$$

Using the fact that $(x+y)^2 \geq 4xy$ we obtain

$$\|f^*_{\lambda,D}(h)\|^2_{F_{3,\nu}(\mathbb{C})} \leq \frac{1}{4\lambda} \sum_{n=1}^{\infty} \alpha_{n-1}(3,\nu)|h_{n-1}|^2 = \frac{1}{4\lambda} \|h\|^2_{F_{3,\nu}(\mathbb{C})}.$$

This proves (ii) and completes the proof of the theorem. □
As in the same way of Theorem 4 we also obtain.
Remark 1. If $\lambda > 0$ and $h \in F_{3,\nu}(\mathbb{C})$, then

$$|Df^*_{\lambda,D}(h)(z)|, |f^*_{\lambda,D}(Dh)(z)| \leq \frac{1}{2\sqrt{\lambda}} \left(I_{3,\nu}\left(|z|^2\right)\right)^{1/2} \|f\|_{F_{3,\nu}(\mathbb{C})}.$$

The extremal function $f^*_{\lambda,D}(h)$ possesses also the following approximation formulas.

Theorem 5. If $\lambda > 0$ and $h \in F_{3,\nu}(\mathbb{C})$, then

$$\lim_{\lambda\to 0^+} \|Df^*_{\lambda,D}(h) - h\|^2_{F_{3,\nu}(\mathbb{C})} = 0,$$

and

$$\lim_{\lambda\to 0^+} Df^*_{\lambda,D}(h)(z) = h(z).$$

Proof. Let $\lambda > 0$ and $h \in F_{3,\nu}(\mathbb{C})$ with $h(z) = \sum_{n=0}^{\infty} h_n z^{3n}$. From (9) we have

$$Df^*_{\lambda,D}(h)(z) - h(z) = \sum_{n=0}^{\infty} \frac{-\lambda\alpha_{n+1}(3,\nu)h_n}{\lambda\alpha_{n+1}(3,\nu) + \alpha_n(3,\nu)} z^{3n}.$$

Therefore

$$\|Df^*_{\lambda,D}(h) - h\|^2_{F_{3,\nu}(\mathbb{C})} = \sum_{n=0}^{\infty} \alpha_n(3,\nu)\left[\frac{\lambda\alpha_{n+1}(3,\nu)|h_n|}{\lambda\alpha_{n+1}(3,\nu) + \alpha_n(3,\nu)}\right]^2$$

Using the fact that

$$\alpha_n(3,\nu)\left[\frac{\lambda\alpha_{n+1}(3,\nu)|h_n|}{\lambda\alpha_{n+1}(3,\nu) + \alpha_n(3,\nu)}\right]^2 \le \alpha_n(3,\nu)|h_n|^2,$$

and

$$\frac{\lambda\alpha_{n+1}(3,\nu)|h_n|}{\lambda\alpha_{n+1}(3,\nu) + \alpha_n(3,\nu)}|z|^{3n} \le |h_n||z|^{3n},$$

we obtain the results from dominated convergence theorem. □

As in the same way of Theorem 5 we also obtain.

Remark 2. If $\lambda > 0$ and $h \in F_{3,\nu}(\mathbb{C})$, then

$$\lim_{\lambda\to 0^+} \|f^*_{\lambda,D}(Dh) - (h - h_0)\|^2_{F_{3,\nu}(\mathbb{C})} = 0,$$

and

$$\lim_{\lambda\to 0^+} f^*_{\lambda,D}(Dh)(z) = h(z) - h(0),$$

where $h_0(z) = h(0)$.

3.2 The integral operator

Let P be the integral operator defined by

$$Pf(z) := \langle f, \Phi_z\rangle_{F_{3,\nu}(\mathbb{C})} = \int_{\mathbb{C}} f(w)\overline{\Phi_z(w)}\,dm_{3,\nu}(w),$$

where

$$\Phi_z(w) = \sum_{n=0}^{\infty} \frac{(\overline{z})^{3n+3} w^{3n}}{\left(\sqrt{n+1}\right)^3 \alpha_n(3,\nu)}, \quad w \in \mathbb{C}.$$

If $f(z) = \sum_{n=0}^{\infty} a_n z^{3n}$, then $Pf(z) = \sum_{n=1}^{\infty} \frac{a_{n-1}}{(\sqrt{n})^3} z^{3n}$.

In this subsection, we determine the extremal function $f^*_{\lambda,P}$ associated to the integral operator P on the space $F_{3,\nu}(\mathbb{C})$.

Theorem 6.

i. The operator P maps continuously from $F_{3,\nu}(\mathbb{C})$ into $F_{3,\nu}(\mathbb{C})$, and

$$\|Pf\|_{F_{3,\nu}(\mathbb{C})} \le \sqrt{27(\nu+1)} \|f\|_{F_{3,\nu}(\mathbb{C})}.$$

ii. For $f \in F_{3,\nu}(\mathbb{C})$ with $f(z) = \sum_{n=0}^{\infty} a_n z^{3n}$, we have

$$P^*f(z) = \sum_{n=0}^{\infty} \frac{\alpha_{n+1}(3,\nu) a_{n+1}}{\left(\sqrt{n+1}\right)^3 \alpha_n(3,\nu)} z^{3n}, \quad P^*Pf(z) = \sum_{n=0}^{\infty} \frac{\alpha_{n+1}(3,\nu) a_n}{(n+1)^3 \alpha_n(3,\nu)} z^{3n}.$$

iii. For any $h \in F_{3,\nu}(\mathbb{C})$ and for any $\lambda > 0$, the problem

$$\inf_{f \in F_{3,\nu}(\mathbb{C})} \left\{ \lambda \|f\|^2_{F_{3,\nu}(\mathbb{C})} + \|Pf - h\|^2_{F_{3,\nu}(\mathbb{C})} \right\}$$

has a unique minimizer given by

$$f^*_{\lambda,P}(h)(z) = \langle h, \Psi_z \rangle_{F_{3,\nu}(\mathbb{C})},$$

where

$$\Psi_z(w) = \sum_{n=1}^{\infty} \frac{(\sqrt{n})^3 (\overline{z})^{3(n-1)} w^{3n}}{\lambda n^3 \alpha_{n-1}(3,\nu) + \alpha_n(3,\nu)}, \quad w \in \mathbb{C}.$$

Proof.

i. If $f \in F_{3,\nu}(\mathbb{C})$ with $f(z) = \sum_{n=0}^{\infty} a_n z^n$, then

$$\|Pf\|^2_{F_{3,\nu}(\mathbb{C})} = \sum_{n=1}^{\infty} \frac{\alpha_n(3,\nu)}{n^3} |a_{n-1}|^2 \le 27(\nu+1) \sum_{n=0}^{\infty} \alpha_n(3,\nu) |a_n|^2 = 27(\nu+1) \|f\|^2_{F_{3,\nu}(\mathbb{C})}.$$

ii. If $f, g \in F_{3,\nu}(\mathbb{C})$ with $f(z) = \sum_{n=0}^{\infty} a_n z^{3n}$ and $g(z) = \sum_{n=0}^{\infty} b_n z^{3n}$, then

$$\langle Pf, g \rangle_{F_{3,\nu}(\mathbb{C})} = \sum_{n=1}^{\infty} \alpha_n(3,\nu) \frac{a_{n-1}}{(\sqrt{n})^3} \overline{b_n} = \sum_{n=0}^{\infty} \alpha_{n+1}(3,\nu) \frac{a_n}{\left(\sqrt{n+1}\right)^3} \overline{b_{n+1}} = \langle f, P^*g \rangle_{F_{3,\nu}(\mathbb{C})},$$

where

$$P^*g(z) = \sum_{n=0}^{\infty} \frac{\alpha_{n+1}(3,\nu)b_{n+1}}{\left(\sqrt{n+1}\right)^3 \alpha_n(3,\nu)} z^{3n}.$$

And therefore

$$P^*Pf(z) = \sum_{n=0}^{\infty} \frac{\alpha_{n+1}(3,\nu)a_n}{(n+1)^3 \alpha_n(3,\nu)} z^{3n}.$$

iii. We put $h(z) = \sum_{n=0}^{\infty} h_n z^{3n}$ and $f^*_{\lambda,P}(h)(z) = \sum_{n=0}^{\infty} c_n z^{3n}$. From (8) we have $(\lambda I + P^*P)f^*_{\lambda,P}(h)(z) = P^*h(z)$. By (ii) we deduce that

$$c_n = \frac{\left(\sqrt{n+1}\right)^3 \alpha_{n+1}(3,\nu)h_{n+1}}{\lambda(n+1)^3\alpha_n(3,\nu) + \alpha_{n+1}(3,\nu)}, \quad n \in \mathbb{N}.$$

Thus

$$f^*_{\lambda,P}(h)(z) = \sum_{n=1}^{\infty} \frac{\left(\sqrt{n}\right)^3 \alpha_n(3,\nu)h_n z^{3n-3}}{\lambda n^3 \alpha_{n-1}(3,\nu) + \alpha_n(3,\nu)} = \langle h, \Psi_z \rangle_{F_{3,\nu}(\mathbb{C})}, \tag{10}$$

where

$$\Psi_z(w) = \sum_{n=1}^{\infty} \frac{\left(\sqrt{n}\right)^3 (\overline{z})^{3n-3} w^{3n}}{\lambda n^3 \alpha_{n-1}(3,\nu) + \alpha_n(3,\nu)}.$$

This completes the proof of the theorem. □

The extremal function $f^*_{\lambda,P}(h)$ possesses the following properties.

Theorem 7. If $\lambda > 0$ and $h \in F_{3,\nu}(\mathbb{C})$, then

$$|f^*_{\lambda,P}(h)(z)| \le \frac{1}{2\sqrt{\lambda}} \left(I_{3,\nu}\left(|z|^2\right)\right)^{1/2} \|h\|_{F_{3,\nu}(\mathbb{C})},$$

$$\|f^*_{\lambda,P}(h)\|_{F_{3,\nu}(\mathbb{C})} \le \frac{1}{2\sqrt{\lambda}} \|h\|_{F_{3,\nu}(\mathbb{C})}.$$

Proof. Let $\lambda > 0$ and $h \in F_{3,\nu}(\mathbb{C})$ with $h(z) = \sum_{n=0}^{\infty} h_n z^{3n}$.

i. From (10) we have

$$|f^*_{\lambda,P}(h)(z)| \le \|\Psi_z\|_{F_{3,\nu}(\mathbb{C})} \|h\|_{F_{3,\nu}(\mathbb{C})}.$$

And by using the fact that $(x+y)^2 \ge 4xy$ we obtain

$$\|\Psi_z\|^2_{F_{3,\nu}(\mathbb{C})} = \sum_{n=1}^{\infty} \frac{n^3 \alpha_n(3,\nu)|z|^{6n-6}}{[\lambda n^3 \alpha_{n-1}(3,\nu) + \alpha_n(3,\nu)]^2} \le \frac{1}{4\lambda} \sum_{n=0}^{\infty} \frac{|z|^{6n}}{\alpha_n(3,\nu)} = \frac{1}{4\lambda} I_{3,\nu}\left(|z|^2\right).$$

This proves (i).
ii. From (10) we have

$$\|f^*_{\lambda,P}(h)\|^2_{F_{3,\nu}(\mathbb{C})} = \sum_{n=0}^{\infty} \alpha_n(3,\nu)\left[\frac{(\sqrt{n+1})^3\alpha_{n+1}(3,\nu)|h_{n+1}|}{\lambda(n+1)^3\alpha_n(3,\nu)+\alpha_{n+1}(3,\nu)}\right]^2.$$

Using the fact that $(x+y)^2 \geq 4xy$ we obtain

$$\|f^*_{\lambda,P}(h)\|^2_{F_{3,\nu}(\mathbb{C})} \leq \frac{1}{4\lambda}\sum_{n=0}^{\infty} \alpha_{n+1}(3,\nu)|h_{n+1}|^2 \leq \frac{1}{4\lambda}\|h\|^2_{F_{3,\nu}(\mathbb{C})}.$$

This proves (ii) and completes the proof of the theorem. □
As in the same way of Theorem 7 we also obtain.
Remark 3. If $\lambda > 0$ and $h \in F_{3,\nu}(\mathbb{C})$, then

$$|Pf^*_{\lambda,P}(h)(z)|, |f^*_{\lambda,P}(Ph)(z)| \leq \frac{3\sqrt{3}}{2}\sqrt{\frac{\nu+1}{\lambda}}\left(I_{3,\nu}\left(|z|^2\right)\right)^{1/2}\|h\|_{F_{3,\nu}(\mathbb{C})}.$$

The extremal function $f^*_{\lambda,P}(h)$ possesses also the following approximation formulas.
Theorem 8. If $\lambda > 0$ and $h \in F_{3,\nu}(\mathbb{C})$, then

$$\lim_{\lambda\to 0^+}\|Pf^*_{\lambda,P}(h) - (h - h_0)\|^2_{F_{3,\nu}(\mathbb{C})} = 0,$$

and

$$\lim_{\lambda\to 0^+} Pf^*_{\lambda,P}(h)(z) = h(z) - h(0).$$

Proof. Let $\lambda > 0$ and $h \in F_{3,\nu}(\mathbb{C})$ with $h(z) = \sum_{n=0}^{\infty} h_n z^{3n}$. From (10) we have

$$Pf^*_{\lambda,P}(h)(z) - (h(z) - h(0)) = \sum_{n=1}^{\infty}\frac{-\lambda n^3\alpha_{n-1}(3,\nu)h_n}{\lambda n^3\alpha_{n-1}(3,\nu)+\alpha_n(3,\nu)}z^{3n}.$$

Therefore

$$\|Pf^*_{\lambda,P}(h) - (h - h_0)\|^2_{F_{3,\nu}(\mathbb{C})} = \sum_{n=1}^{\infty}\alpha_n(3,\nu)\left[\frac{\lambda n^3\alpha_{n-1}(3,\nu)|h_n|}{\lambda n^3\alpha_{n-1}(3,\nu)+\alpha_n(3,\nu)}\right]^2$$

Using the fact that

$$\alpha_n(3,\nu)\left[\frac{\lambda n^3\alpha_{n-1}(3,\nu)|h_n|}{\lambda n^3\alpha_{n-1}(3,\nu)+\alpha_n(3,\nu)}\right]^2 \leq \alpha_n(3,\nu)|h_n|^2,$$

and

$$\frac{\lambda n^3\alpha_{n-1}(3,\nu)|h_n|}{\lambda n^3\alpha_{n-1}(3,\nu)+\alpha_n(3,\nu)}|z|^{3n} \leq |h_n||z|^{3n},$$

we obtain the results from dominated convergence theorem. □

As in the same of Theorem 8 we also obtain.

Remark 4. If $\lambda > 0$ and $h \in F_{3,\nu}(\mathbb{C})$, then

$$\lim_{\lambda \to 0^+} \|f^*_{\lambda,P}(Ph) - h\|^2_{F_{3,\nu}(\mathbb{C})} = 0,$$

and

$$\lim_{\lambda \to 0^+} f^*_{\lambda,P}(Ph)(z) = h(z).$$

4. Conclusion and perspectives

Bargmann [1] in 1961 introduced the classical Fock space $F(\mathbb{C})$ and Cholewinsky [2] in 1984 introduced the generalized Fock space $F_{2,\nu}(\mathbb{C})$. The Bessel-type Fock space $F_{3,\nu}(\mathbb{C})$ introduced in this work generalizes these analytic spaces. We are studied the Tikhonov regularization problem associated to this Hilbert space, and we are established the extremal function for this problem. Finally, in a future paper we have the idea to study the Bessel-type Segal-Bargmann transform, in which we will prove inversion formula, Plancherel formula and some uncertainty inequalities for this transform.

Conflicts of interest

The author declares that there is no conflict of interests regarding the publication of this paper.

Author details

Fethi Soltani[1,2]

1 Faculté des Sciences de Tunis, Laboratoire d'Analyse Mathématique et Applications LR11ES11, Université de Tunis El Manar, Tunis, Tunisia

2 Ecole Nationale d'Ingénieurs de Carthage, Université de Carthage, Tunis, Tunisia

*Address all correspondence to: fethi.soltani@fst.utm.tn

References

[1] Bargmann V. On a Hilbert space of analytic functions and an associated integral transform part I. Communications on Pure and Applied Mathematics. 1961;**14**:187-214

[2] Cholewinski FM. Generalized Fock spaces and associated operators. SIAM Journal on Mathematical Analysis. 1984; **15**:177-202

[3] Erdely A et al. Higher Transcendental Functions. Vol. 2. New York: McGraw-Hill; 1953

[4] Baccar C, Omri S, Rachdi LT. Fock spaces connected with spherical mean operator and associated operators. Mediterranean Journal of Mathematics. 2009;**6**(1):1-25

[5] Berger CA, Coburn LA. Toeplitz operators on the Segal-Bargmann space. Transactions of the American Mathematical Society. 1987;**301**:813-829

[6] Nemri A, Soltani F, Abd-alla AN. New studies on the Fock space associated to the generalized Airy operator and applications. Complex Analysis and Operator Theory. 2018;**12**(7):1549-1566

[7] Sifi M, Soltani F. Generalized Fock spaces and Weyl relations for the Dunkl kernel on the real line. Journal of Mathematical Analysis and Applications. 2002;**270**:92-106

[8] Soltani F. Fock-type spaces associated to higher-order Bessel operator. Integral Transforms and Special Functions. 2018; **29**(7):514-526

[9] Kiryakova V. Obrechkoff integral transform and hyper-Bessel operators via *G*-function and fractional calculus approach. Global Journal of Pure Applied Mathematics. 2005;**1**(3):321-341

[10] Widder DV. The Airy transform. American Mathematical Monthly. 1979; **86**:271-277

[11] Cholewinski FM, Reneke JA. The generalized Airy diffusion equation. Electronic Journal of Differential Equations. 2003;**87**:1-64

[12] Fitouhi A, Mahmoud NH, Ahmed Mahmoud SA. Polynomial expansions for solutions of higher-order Bessel heat equations. Journal of Mathematical Analysis and Applications. 1997;**206**: 155-167

[13] Fitouhi A, Ben Hammouda MS, Binous W. On a third singular differential operator and transmutation. Far East Journal of Mathematical Sciences. 2006;**21**(3):303-329

[14] Dimovski I, Kiryakova V. The Obrechkoff, integral transform: properties and relation to a generalized fractional calculus. Numerical Functional Analysis and Optimization. 2000;**21**(1–2):121-144

[15] Ben Hammouda MS, Nemri A. Polynomial expansions for solutions of higher-order Besse heat equation in quantum calculus. Fractional Calculus and Applied Analysis. 2007; **10**(1):39-58

[16] Matsuura T, Saitoh S, Trong D. Inversion formulas in heat conduction multidimensional spaces. Journal of Inverse and Ill-Posed Problems. 2005;**13**: 479-493

[17] Matsuura T, Saitoh S. Analytical and numerical inversion formulas in the Gaussian convolution by using the Paley-Wiener spaces. Applicable Analysis. 2006;**85**:901-915

[18] Soltani F. Extremal functions on Sturm-Liouville hypergroups. Complex Analysis and Operator Theory. 2014; **8**(1):311-325

[19] Soltani F. Extremal functions on Sobolev-Dunkl spaces. Integral Transforms and Special Functions. 2013; **24**(7):582-595

[20] Soltani F. Multiplier operators and extremal functions related to the dual Dunkl-Sonine operator. Acta Mathematica Scientia. 2013;**33B**(2): 430-442

[21] Paulsen VI. An Introduction to the Theory of Reproducing Kernel Hilbert Spaces. Cambridge: Cambridge University Press; 2016

[22] Folland G. Harmonic Analysis on Phase Space. Princeton, New Jersey: Princeton University Press; 1989

[23] Gröchenig K. Foundations of Time-frequency Analysis. Boston: Birkhäuser; 2001

[24] Saitoh S. Best approximation, Tikhonov regularization and reproducing kernels. Kodai Mathematical Journal. 2005;**28**:359-367

[25] Saitoh S. Theory of reproducing kernels: Applications to approximate solutions of bounded linear operator equations on Hilbert spaces. In book: Selected papers on analysis and differential equations. Amer. Math. Soc. Transl. Series 2. 2010;**230**:107-134

[26] Saitoh S, Sawano Y. Theory of reproducing kernels and applications. Developements in Mathematics. 2016; **44**:4

Chapter 7

On Principal Parts-Extension for a Noether Operator A

Abdourahman Haman Adji

Abstract

The main purpose of this work is to realize and establish the extension of a noether operator A defined by a third kind singular integral equation. The extended operator of the initial operator A is noted $\hat{A}$ and acting from a new specific functional space constructed denoted $V_m = C^1_{-1}[-1, 1] \oplus \left\{\sum_{k=1}^{m} \alpha_k F.p\frac{1}{x^k}\right\}$, into the same previous space $C_0^{\{P\}}[-1, 1]$. We investigate the noetherity of the extended initial noether operator noted $\hat{A}$, when we realized the extension taking the unknown function $\varphi(x)$ from the space V_m. The index of such noether operator $\chi\left(\hat{A}\right)$ is calculated and therefore, the conditions of the noetherity nature of the extended operator are established.

Keywords: integral equation of the third kind, deficient numbers, index, Taylor derivative, noether operator, fundamental functions, singular operator

1. Introduction

The construction of noether theory for some integrodifferential operators defined by linear third-kind integral equations in some specific functional spaces is well known and still interests many scientists around the world. Various scientific works dedicated to the noetherity of integrodifferential operators have been published by many researchers investigating such topics.

In our previous works, while constructing noether theory for integrodifferential operators defined by the third kind integral equations, we approached the question of the solvability of linear integral equations of the third kind $a(x)I + K$, where K is an integral operator and $a(x)$ is a given function vanishing on some set of points. That was illustrated in many works (see for example, papers [1–12] and scientific collections [13, 14]. We also recall that such equations arise frequently in various applications and by their particular nature are connected with singularities of the function $a(x)$. In connexion with integral equations of the third kind, we can note that they have been investigated at the beginning of the centenary. In addition, there is a particular interest linked to them in relation to the requirement of transport theory without forgetting the theory of elliptical-hyperbolic type equations. In most scientific research works, the solution of the integral equation of the third kind seems to be the space of continuous functions on the closed interval $[a, b]$.

Specific approaches are needed when constructing noether theory for integrodifferential operators and, once we succeed to establish the noetherity of the considered operator, we can always set to the problem of the investigation of the extension of the studied already noether operator. For some details see [15–18].

The illustration of the great importance of the construction and establishment of noetherity of some integrodifferential operators defined by third-kind integral equations is clearly presented through the works of many scientists. For full details on such investigations and results, we can refer to works such as [3–5, 15, 16, 19–26]. Let us recall that among many others, Bart G.R, Warnock R.L., Roghozin V., Raslambekov V. S., and Gobbassov N.S., respectively, in their scientific works, had constructed noether theory for some integrodifferential operators defined similarly but having some specificities on the considered integral equation. They have realized some cases of extension and established noetherity of the considered extended operator.

We recall that the noetherity of the initial operator $(A\varphi)(x) = x^p\varphi'(x) + \int_{-1}^{1}K(x,t)\varphi(t)dt = f(x); x \in [-1, 1]$, with $\varphi \in C^1_{-1}[-1, 1], f(x) \in C_0^{\{P\}}[-1, 1]$ and $K(x,t) \in C_0^{\{P\}}[-1, 1]$X C $[-1, 1]$ is completely established in [4, 5].

E.Tompé and al, in their recent published article titled "On Delta-Extension for a Noether Operator" have realized the extension of the initial operator $(A\varphi)(x) = x^p\varphi'(x) + \int_{-1}^{1}K(x,t)\varphi(t)dt = f(x); x \in [-1, 1]$, when the unknown function rather than $\varphi \in C^1_{-1}[-1, 1]$, was took from the space $D_m = C^1_{-1}[-1, 1] \oplus \left\{\sum_{k=0}^{m}\alpha_k\delta^{\{k\}}(x)\right\}$ with the conditions $0 \le m \le p-2$. For full details see [22].

Just recently their paper titled "Noetherity of a Dirac Delta-Extension for a Noether Operator" was published in the *International Journal of Theoretical and Applied Mathematics*. Vol. 8, No. 3, 2022, pp. 51–57. doi: 10.11648/j.ijtam.20220803.11, Abdourahman, Ecclésiaste Tompé Weimbapou, and Emmanuel Kengne completely covered the investigation of the noetherity of the extended operator $\overline{A}$ of the initial operator $(A\varphi)(x) = x^p\varphi'(x) + \int_{-1}^{1}K(x,t)\varphi(t)dt = f(x); x \in [-1, 1]$, previously started, when also, at this time the unknown function rather than $\varphi \in C^1_{-1}[-1, 1]$has been taken as following
$\varphi \in D_m = C^1_{-1}[-1, 1] \oplus \left\{\sum_{k=0}^{n}\alpha_k\delta^{\{k\}}(x)\right\}$ with supplementary condition$m > p-2$.

The noetherity, in both two cases, investigated, of the extended operator noted $\overline{A}$ was established and its index $\chi(\overline{A})$ is calculated.

Following such previous research cited and other works are done by many scientists, related to the realization of various types of extensions of noether operators, we are conducting the work to realize a particular type of extension when we add, at this time, functions from the space of principal-parts values of the following indicated form $\left\{\sum_{k=1}^{m}\alpha_k F.p\frac{1}{x^k}\right\}$.

Namely here in this paper, we realize the extension of the following noether operator defined by the third kind singular integral equation

$$(A\varphi)(x) = x^p\varphi'(x) + \int_{-1}^{1}K(x,t)\varphi(t)dt = f(x); x \in [-1, 1] \tag{1}$$

where $\varphi \in C^1_{-1}[-1, 1], f(x) \in C_0^{\{P\}}[-1, 1]$ and $K(x,t) \in C_0^{\{P\}}[-1, 1]$X C $[-1, 1]$ with principal parts functions, i.e., $\varphi \in V_m = C^1_{-1}[-1, 1] \oplus \left\{\sum_{k=1}^{m}\alpha_k F.p\frac{1}{x^k}\right\}$ and next, we establish the noetherity of the extended operator.

The structure of this chapter is the following: Section 2 is devoted to some fundamental well-known notions and concepts of noether theory, Fredholm third kind integral equation, Taylor derivatives, associated spaces, and associated operators. Section 3 presenting the main results of the chapter deals properly with the realization of the extension of the operator A when taking the unknown function from the space V_m. Lastly, after making a small important remark, we conclude our chapter in Section 4, followed by some recommendations for the follow-up or future scientific works to undertake, as stated in Section 5.

2. Preliminaries

Before presenting our main results in full detail, the following definitions and concepts well known from the noether theory of operators are required for the realization of this research. We also recall the notions of Taylor derivatives and linear Fredholm integral equation of the third kind, widely studied in many works done by different authors among many of them Bart GR, Sukavanam N, Shulaia D.A, Gobbassov N.S. See [3, 27–31] for more details.

First of all, let us move to the following concept.

A. Noether operator

Definition 1. Let X, Y be Banach spaces, $A \in l(X, Y)$ a linear operator. The quotient space$coker\ A = Y/imA$ is called the cokernel of the operator A. The dimensions $\alpha(A) = \dim kerA$,$\beta(A) = \dim cokerA$ are called the nullity and the deficiency of the operator A, respectively. If at least one of the numbers $\alpha(A)$ or $\beta(A)$ is finite, then the difference $Ind\ A = \alpha(A) - \beta(A)$ is called the index of the operator A.

Definition 2. Let X, Y be Banach spaces, $A \in l(X, Y)$ is said to be normally solvable if it possesses the following property: The equation $Ax = y\ (y \in Y)(y \in Y)$ has at least one solution $x \in D(A)$ ($D(A)$ is the domain of A) if and only if $<y,f> = 0 \forall f \in (im\ A)^{\perp}$ holds.

We recall that by the definition of the adjoint operator $(im\ A)^{\perp} = \ker A^*$ and it's proven in [4] that The operator A is normally solvable if and only if its image space imA is closed.

Definition 3. A closed normally solvable operator A is called a Noether operator if its index is finite.

By the way, we briefly review these important notions of Taylor derivatives which are widely used when constructing noether theory of the considered operator A.

Definition 4. A Continuous function $\varphi(x)\ \epsilon\ C[-1, 1]$ admits at the point $x = 0$ Taylor derivative up to the order $p \in \mathbb{N}$ if there exists recurrently for $k = 1,2, \dots ,p$, the following limits:

$$\varphi^{\{k\}}(0) = k! \lim_{x \to 0} x^{-k} \left[\varphi(x) - \sum_{j=0}^{k-1} \frac{\varphi^{\{j\}}(0)}{j!} x^j \right] \tag{2}$$

The class of such functions $\varphi(x)$ is denoted $C_0^{\{p\}}[-1,\ 1]$.

Next, let us move to the following part.

Let $C^m[-1,\ 1], m \in \mathbb{Z}_+$, noted the Banach space of continuous functions on $[-1,\ 1]$, having continuous derivatives up to order m, for which the norm is defined as follows:

$$\|\varphi(x)\|_{C^m[-1,\ 1]} = \sum_{j=0}^{m} \max_{-1\le x\le 1} \left|\varphi^{(j)}(x)\right| \tag{3}$$

Therefore, we can consider $\varphi^{\{k\}}(0)$ are defined for all $k = 1,2, \dots ,p$.

We define $C_0^{\{p\}}[-1,\ 1]$ as a subspace of continuous functions, having finite Taylor derivatives up to order $p \in \mathbb{Z}_+$; and when $p = 0$, we put $\left(C_0^{\{p\}}[-1,\ 1] = C_0^{\{0\}}[-1,\ 1] = C[-1,\ 1]\right)$.

Let us also define a linear operator N^k on the space $C_0^{\{p\}}[-1,\ 1]$ by the formula:

$$\left(N^k\varphi\right)(x) = \frac{\varphi(x) - \sum_{j=0}^{k-1} \frac{\varphi^{\{j\}}(0)}{j!} x^j}{x^k}, k = 1,2, \dots ,p \tag{4}$$

One can easily verify the property $N^k = N^{k_1} N^{k-k_1}, 0 \le k_1 \le k, k, k_1 \in \mathbb{Z}_+$, where we put $N^0 = I$.

Definition 5. The operator N^p is called the characteristical operator of the space $C_0^{\{p\}}[-1,\ 1]$.

Remark: The sense of the previous definition can be seen from the verification of the following lemma and also for more details see [23, 25, 26].

Lemma 2.1. A function $\varphi(x)$ belongs to $C_0^{\{p\}}[-1,\ 1]$ if and only if, the following representation

$$\varphi(x) = x^p \phi(x) + \sum_{k=0}^{p-1} \alpha_k x^k \tag{5}$$

holds with the function $\phi(x) \in C[-1,\ 1]$, and α_k being constants.

To prove Lemma 2.1 it is enough to observe that (5) implies that the Taylor derivatives of $\varphi(x)$ up to the order p exists, and more $\varphi^{\{k\}}(0) = k!\alpha_k, k = 0,1,2, \dots ,p-1, \varphi^{\{0\}}(0) = p!\phi(0)$ with $\phi(x) = \left(N^k\varphi\right)(x)$. Conversely, if $\varphi(x)$ belongs to $C_0^{\{p\}}[-1,\ 1]$, and we define $\phi(x) = \left(N^k\varphi\right)(x)$ with $\alpha_k = \frac{\varphi^{\{k\}}(0)}{k!}, k = 0,1,2, \dots ,p-1$, then the representation (5) holds. From Lemma 2.1, it follows that for $\varphi(x) \in C_0^{\{p\}}[-1,\ 1]$ the inequality

$$\varphi(x) = x^p \left(N^k\varphi\right)(x) + \sum_{k=0}^{p-1} \frac{\varphi^{\{k\}}(0)}{k!} x^k, \tag{6}$$

is valid.

Consequently, the linear operator N^p establishes a relation between the spaces $C_0^{\{p\}}[-1, 1]$ and $C[-1, 1]$. The space $C_0^{\{p\}}[-1, 1]$ with the norm

$$\|\varphi\|_{C_0^{\{p\}}[-1, 1]} = \|N^p\varphi\|_{C[-1, 1]} + \sum_{k=0}^{p-1} |\varphi^{\{k\}}(0)|, \tag{7}$$

becomes a Banach space one.

Let note also that we can define the previous norm in the following way:

$$\|\varphi\|_{C_0^{\{p\}}[-1, 1]} = \|N^p\varphi\|_{C[-1, 1]} + \sum_{k=0}^{p-1} |\alpha_k| = \|\phi(x)\|_{C[-1, 1]} + \sum_{k=0}^{p-1} |\alpha_k|. \tag{8}$$

Sometimes it is comfortable and suitable to consider as the norm in the space $C_0^{\{p\}}[-1, 1]$ the equivalent norm is defined as follows:

$$\|\varphi\|_{C_0^{\{p\}}[-1, 1]} = \sum_{j=0}^{p} \|N^j\varphi\|_{C[-1, 1]} \tag{9}$$

We can also note a very useful and clearly helpful next inequality:

$$\|\varphi\|_{C[-1, 1]} \leq \|N^p\varphi\|_{C[-1, 1]} + \sum_{j=0}^{p-1} |\varphi^{\{j\}}(0)| = \|\varphi\|_{C_0^{\{p\}}[-1, 1]} \tag{10}$$

Therefore, it is obvious to see that

$$\|\varphi\|_{C[-1, 1]} \leq \|\varphi\|_{C_0^{\{p\}}[-1, 1]}. \tag{11}$$

Finally, note that from definition 2.1, we can follow fact that if $\varphi(x) \in C[-1, 1]$, then $x^p\varphi(x) \in C_0^{\{p\}}[-1, 1]$. This assertion may be generalized as follows:

Lemma 2.2. Let $p \in \mathbb{N}, s \in \mathbb{Z}_+$. If $\varphi(x) \in C_0^{\{s\}}[-1, 1]$ then, $x^p\varphi(x) \in C_0^{\{p+s\}}[-1, 1]$, and the formula holds

$$(x^p\varphi(x))^{\{j\}}(0) = \begin{cases} 0, j = 0, 1, \dots, p-1, \\ \dfrac{j!}{(j-p)!}\varphi^{\{j-p\}}(0), j = p, \dots, p+s. \end{cases} \tag{12}$$

Proof. Note that a stronger assumption on the function $\varphi(x)$, such that $\varphi(x) \in C_0^{\{p+s\}}[-1, 1]$ would allow us to easily prove the lemma just by applying the Leibniz formula.

For $s = 0$ the statement has already been proved above, so $x^p\varphi(x) \in C_0^{\{p\}}[-1, 1]$, and $(x^p\varphi(x))^{\{j\}}(0) = 0, j = 0, \dots, p-1$ and

$(x^p\varphi(x))^{\{p\}}(0) = p!\varphi(0)$. Now let us prove that $(x^p\varphi(x))^{\{j\}}(0) = \frac{j!}{(j-p)!}\varphi^{\{j-p\}}(0), j = p+1, \dots, p+s$. Since the derivatives are defined recurrently, and (12) is true for $j = p$, then it is sufficient to verify the passage from j to $j+1$. We have:

$$(x^p\varphi(x))^{\{j+1\}}(0) = (j+1)!\lim_{x\to 0}\frac{x^p\varphi(x) - \sum_{l=p}^{j}\frac{x^l}{(l-p)!}\varphi^{\{l-p\}}(0)}{x^{j+1}} \tag{13}$$

$$= (j+1)!\lim_{x\to 0}\frac{\varphi(x) - \sum_{l=0}^{j-p}\frac{x^l\varphi^{\{l\}}(0)}{l!}}{x^{j+1-p}} = \frac{(j+1)!}{(j+1-p)!}\varphi^{\{j+1-p\}}(0). \tag{14}$$

Lemmas 2.1 and 2.2 imply the next important lemma.

Lemma 2.3.

Let $f(x) \in C_0^{\{p\}}[-1, 1], p \in \mathbb{N}$ and $f(0) = \dots\dots = f^{\{r-1\}}(0) = 0, 1 \le r \le p$. Then $\frac{f(x)}{x^r} \in C_0^{\{p-s\}}[-1, 1]$.

We say that the kernel $k(x, t) \in C_0^{\{P\}}[-1, 1]\, XC[-1, 1]$, if and only if $k(x, t) \in C[-1, 1]X\, C[-1, 1]$ and admits Taylor derivatives according to the variable x at the point $(0,t)$ whatever $t \in [-1, 1]$.

B. Associated spaces and associated operators

Instead of talking about adjoint operators when establishing the noetherity of an operator, we can note that also noether property of an operator may depend on the concept of associated operators and associated spaces. Therefore, we start by recalling these two important concepts and we give some associated spaces that we are going to use later within the work.

Definition 6.

The Banach space $E' \subset E^*$ is called associated space with a Banach space E, if

$$|<f, \varphi>| \le c\|f\|_{E'}\|\varphi\|_E \forall\varphi \in E, f \in E'. \tag{15}$$

Definition 6. Let E_j, (j = 1,2) be Banach spaces and E_j' their associated spaces, operators $A \in l(E_1, E_2)$ and $A' \in l(E_1', E_2')$ are associated if and only if:

$$(A'f, \varphi) = (f, A\varphi)\forall f \in E_2' \text{ and } \varphi \in E_1. \tag{16}$$

The following important result gives noether property via an associated operator.

Lemma 1. Let E_j, j = 1,2 be Banach spaces, E_j' their associated spaces, $A \in l(E_1, E_2)$ and $A' \in l(E_1', E_2')$ are associated with noether operators, we have $\chi(A) = -\chi(A')$ (where χ means the index), and for the solvability of equation

$A\varphi = f$ it's necessary and sufficient that $(f, \psi) = 0$ for all solutions of the associated homogenous equation $A'\psi = 0$.

We finish these reminders with two very important results that define the associated spaces of spaces that we will use later.

Lemma 2. Space $C^1_{x_0}[-1, 1]$ is associated with space $C[-1, 1]$ where $C^1_{x_0}[-1, 1]$ means the space of functions $\varphi \in C[-1, 1]$ satisfying $\varphi(x_0) = 0$.

Proof. Let $f \in C^1_{x_0}[-1, 1]$ and $\varphi \in C[-1, 1]$. Then we have:

$$|<\mathrm{f}, \varphi>| = \left|\int_{-1}^{1} \mathrm{f(x)}\varphi(\mathrm{x})\mathrm{dx}\right| \leq 2\max{}_{-1\leq x\leq 1}|\mathrm{f(x)}|.\max{}_{-1\leq x\leq 1}|\varphi(\mathrm{x})|.\blacksquare \tag{17}$$

Let us also recall the definition of the space of generalized functions P^1 given in [12, 22, 23].

Definition 7. Through P^1 we denote the space of distributions ψ on the space of test functions $C^{\{p\}}_{-1}[-1, 1]$ such that:

$\psi(x) = \frac{z(x)}{x^p} + \sum_{k=0}^{p-1} \beta_k \delta^{\{k\}}(x)$ where $(x) \in C^{\{p\}}_{-1}[-1, 1] \cap C^1_{-1}[-1, 1]$, β_k are arbitrary constants, $\delta^{\{k\}}(x)$is the $k - th$ Taylor derivative of the Dirac-delta function defined by:

$$\left(\delta^{\{k\}}(x), \varphi(x)\right) = \int_{-\infty}^{+\infty} \delta^{\{k\}}\varphi(x)dx = (-1)^k \varphi^{\{k\}}(0). \tag{18}$$

In the space P^1 let us introduce the norm in the following way:

$$\|\psi\|_{P^1} = \|z\|_{C^{\{p\}}_{-1}[-1, 1]} + \|z\|_{C^1[-1, 1]} + \sum_{k=0}^{p-1} |\beta_k| \tag{19}$$

with this norm, it was proved in [12, 22, 23] that the space P^1 is a Banach space.

Lemma 3. The space P^1 is a Banach space associated with the space $C^{\{p\}}_{-1}[-1, 1]$.

Proof. Similar to previous proof and for more details see also [5, 22].

Definition 8. An equation of the form

$$\mathrm{A_n}\varphi = \mathrm{f}, \tag{20}$$

where f is a given function of the variable $x \in [a, b]$ and φ the unknown function of $x \in [a, b]$ when the operator A_n is defined by

$$\mathrm{A_n}\varphi = \mathrm{g_n(x)}\varphi(\mathrm{x}) - \int_{a}^{b} \mathrm{k(x, t)}\varphi(\mathrm{t})\mathrm{dt} \tag{21}$$

is called a linear Fredholm integral equation of the third kind.

In this case, $g_n(x) = \prod_{k=1}^{n}(x - x_k)$ is a given function of the variable $x \in [a, b]$ with $x_k \in]a, b[$ and $k(x, t)$ is a given function of variables $(x, t) \in [a, b]X[a, b]$.

Related to this notion of linear Fredholm integral equation of the third kind with full details also can be found in [5, 22].

Now, we can move to the presentation of the general results of our investigation stated in the following section.

3. Main results

Before we state the content of our results within the whole investigation, let's also give some needed definitions related to the main space and to singular integral for functions we used in the work.

Definition 9. We denote through $V_m = C^1_{-1}[-1, 1] \oplus \left\{\sum_{k=1}^{m} \alpha_k F.p.\left(\frac{1}{x^k}\right)\right\}$ the space of all functions presented as follows:

$$\varphi(x) = \varphi_0(x) + \sum_{k=1}^{m} \alpha_k F.p.\left(\frac{1}{x^k}\right) \tag{22}$$

where $\varphi_0(x) \in C[-1, 1]$ and $\varphi_0(-1) = 0$ with the natural norm

$$\|\varphi\| = \|\varphi_0\|_{C[-1, 1]} + \sum_{k=1}^{m} |\alpha_k|.. \tag{23}$$

Next, we move to the following important concept.

A. Singular integral for functions from the space $C_0^{\{p\}}[-1, 1]$.

If the function $g(x)$ has feature (singularity) in $x = 0$, then we say that $\int_{-1}^{1} g(x)dx$ exists in the sense of Hadamard if it is true the following representation:

$$\int_{-1}^{-\varepsilon} g(x)dx + \int_{\varepsilon}^{1} g(x)dx = a + \sum_{k=1}^{l} a_k \varepsilon^{-k} + a_{l+1} ln\frac{1}{\varepsilon} + 0(\varepsilon), \varepsilon \to 0. \tag{24}$$

In this case, we put $F.p.\int_{-1}^{1} g(x)dx = a$, i.e., it remains the finite parts. Note that under the definition of convergence by Hadamard, we often take $a_{l+1} = 0$, but we do not exclude that possibility as this can allow us to consider the convergence $(V.p.)$ in the sense of Cauchy principal part as a particular case of convergence in the sense of Hadamard.

Now let $\varphi(x) \in C_0^{\{p\}}[-1, 1], p \in \mathbb{N}$ and consider $\int_{-1}^{1} \frac{\varphi(x)}{x^p} dx, p \in \mathbb{N}$.

Lemma 4. Let $\varphi(x) \in C_0^{\{p\}}[-1, 1], p \in \mathbb{N}$. Then it takes place the following relationships:

$$\begin{cases} \text{F.p.}\int_{-1}^{1} \frac{\varphi(x)}{x^p} dx = \int_{-1}^{1} (\text{N}^\text{p}\varphi)(x)dx + \sum_{k=0}^{p-2} \frac{\varphi^{\{k\}}(0)\left(1-(-1)^{k-p+1}\right)}{k-p+1} \\ \text{under } p \geq 2, \\ \text{and} \\ \text{F.p.}\int_{-1}^{1} \frac{\varphi(x)}{x} dx = V.p.\int_{-1}^{1} \frac{\varphi(x)}{x} dx = \int_{-1}^{1} (\text{N}\varphi)(x)dx. \end{cases} \tag{25}$$

Proof. For the proof, we note that by virtue of lemma 2.1 we have $\varphi(x) = x^p(\text{N}^\text{p}\varphi)(x) + \sum_{k=0}^{p-1} \frac{\varphi^{\{k\}}(0)}{k!} x^k$ and next, it remains to note that

$$\begin{cases} \int_{-1}^{-\varepsilon} \frac{dx}{x^{p-k}} + \int_{\varepsilon}^{1} \frac{dx}{x^{p-k}} = \frac{1-(-1)^{k-p+1}}{k-p+1} + \frac{-\varepsilon^{k-p+1} + (-\varepsilon)^{k-p+1}}{k-p+1} \\ \quad \text{when } k \neq p-1 \\ \text{and} \\ \text{and} \int_{-1}^{-\varepsilon} \frac{dx}{x} + \int_{\varepsilon}^{1} \frac{dx}{x} = 0. \end{cases} \tag{26}$$

Consequently,

$$\text{F.p.}\int_{-1}^{1} \frac{dx}{x^{p-k}} = \frac{1-(-1)^{k-p+1}}{k-p+1}, k = 0,1, \ldots \ldots, p-2. \tag{27}$$

That is leading us to the first assertion.

Analogously we can prove the second assertion.

As previously indicated, we note through $\hat{A}$ the extension of the operator A onto the space $C_{-1}^1[-1, 1]$.

We will also note it in the following way:

$$F.p\left(\frac{1}{x^k}\right) = P\left(\frac{1}{x^k}\right) \tag{28}$$

and use the properties

$$\frac{d^l}{dx^l} P\left(\frac{1}{x^k}\right) = P\left(\frac{d^l}{dx^l} \frac{1}{x^k}\right). \tag{29}$$

$$x^l P\left(\frac{1}{x^k}\right) = \begin{cases} x^{l-k}, & l > k \\ 1, & l = k \\ P\left(\frac{1}{x^{k-l}}\right), & l < k. \end{cases} \tag{30}$$

In the following part, we consider our operator defined by the integral equation in the case to be investigated.

B. Integral equation in the case $m = p - 1$

Let $\hat{A}-$ be the extension of the operator A defined by the Eq. (1) and A' is the associated operator to A. Let us explain under which conditions the operators $\hat{A}$ and $\hat{A}'$ are at least formally associated operators.

Let $m = p - 1$. So that we have immediately considered computations:

$$\varphi(x) = \varphi_0(x) + \sum_{k=1}^{m} \alpha_k P\left(\frac{1}{x^k}\right) = \varphi_0(x) + \sum_{k=1}^{p-1} \alpha_k P\left(\frac{1}{x^k}\right). \tag{31}$$

First of all, we calculate $\left(\hat{A}\varphi, \Psi\right)$. Then we have

$$\begin{aligned}\left(\hat{A}\varphi, \Psi\right) &= \left(x^p \varphi'_0(x) - \sum_{k=1}^{p-1} k\alpha_k x^{p-k-1} + \int_{-1}^{1} K(x,t)\varphi_0(t)dt + \sum_{k=1}^{m} \alpha_k \int_{-1}^{1} \frac{K(x,t)}{t^k} dt, \right.\\ &\left.\frac{z(x)}{x^p} + \sum_{n=0}^{p-1} \beta_n \delta^{\{n\}}(x)\right)\\ &= \left(\varphi'_0(x), z(x)\right) - \sum_{k=1}^{p-1} k\alpha_k \int_{-1}^{1} \frac{z(x)}{x^{k+1}} dx + \left(\int_{-1}^{1} K(x, t)\varphi_0(t)dt, \frac{z(x)}{x^p}\right)\\ &+ \sum_{k=1}^{p-1} \alpha_k \left(\int_{-1}^{1} \frac{K(x, t)}{t^k} dt, \frac{z(x)}{x^p}\right) + \sum_{n=0}^{p-1} (-1)^n \beta_n \int_{-1}^{1} K_1^{\{n\}}(0, t)\varphi_0(t)dt\\ &+ \sum_{k=1}^{p-1} \alpha_k \sum_{n=0}^{p-1} (-1)^n \beta_n \int_{-1}^{1} \frac{K_1^{\{n\}}(0,t)}{t^k} dt + \left(-\sum_{k=1}^{p-1} k\alpha_k x^{p-k-1}, \sum_{n=0}^{p-1} \beta_n \delta^{\{n\}}(x)\right).\end{aligned} \tag{32}$$

Next, let us also separately calculate on the other side the following expression:

$$\begin{aligned}\left(\hat{A}\varphi, \Psi\right) &= \left(x^p \varphi'_0(x) - \sum_{k=1}^{p-1} k\alpha_k x^{p-k-1} + \int_{-1}^{1} K(x,t)\varphi_0(t)dt + \sum_{k=1}^{m} \alpha_k \int_{-1}^{1} \frac{K(x,t)}{t^k} dt, \right.\\ &\left.\frac{z(x)}{x^p} + \sum_{n=0}^{p-1} \beta_n \delta^{\{n\}}(x)\right)\\ &= \left(\varphi'_0(x), z(x)\right) - \sum_{k=1}^{p-1} k\alpha_k \int_{-1}^{1} \frac{z(x)}{x^{k+1}} dx + \left(\int_{-1}^{1} K(x, t)\varphi_0(t)dt, \frac{z(x)}{x^p}\right)\\ &+ \sum_{k=1}^{p-1} \alpha_k \left(\int_{-1}^{1} \frac{K(x, t)}{t^k} dt, \frac{z(x)}{x^p}\right) + \sum_{n=0}^{p-1} (-1)^n \beta_n \int_{-1}^{1} K_1^{\{n\}}(0, t)\varphi_0(t)dt\\ &+ \sum_{k=1}^{p-1} \alpha_k \sum_{n=0}^{p-1} (-1)^n \beta_n \int_{-1}^{1} \frac{K_1^{\{n\}}(0,t)}{t^k} dt + \left(-\sum_{k=1}^{p-1} k\alpha_k x^{p-k-1}, \sum_{n=0}^{p-1} \beta_n \delta^{\{n\}}(x)\right).\end{aligned} \tag{33}$$

We rewrite this term in the form of a sum and we obtain definitively the equation as follows:

$$\left(\sum_{n=0}^{p-1} \beta_n \delta^{\{n\}}(x),\ -\sum_{k=1}^{p-1} k\alpha_k x^{p-k-1}\right) = -\sum_{k=1}^{p-1} (-1)^{k-1} \alpha_{p-k}\beta_{k-1}(p-k)(k-1)! \tag{34}$$

On the other side, we compute also the following needed expression:

$$\begin{aligned}\left(\varphi, \hat{A}'\Psi\right) &= \left(\varphi_0(x) + \sum_{k=1}^{p-1} \alpha_k \frac{1}{x^k},\ -(x^p\Psi)' + \int_{-1}^{1} K(t,\ x)\Psi(t)dt\right)\\ &= \left(\varphi_0(x) + \sum_{k=1}^{p-1} \alpha_k \frac{1}{x^k},\ -z'(x) + \int_{-1}^{1} \frac{K(x,\ t)}{t^p} z(t)dt + \sum_{n=0}^{p-1} (-1)^n \beta_n K_1^{\{n\}}(0,\ x)\right)\\ &= (\varphi_0(x),\ -z'(x)) - \sum_{k=1}^{p-1} \alpha_k \left(z'(x),\ \frac{1}{x^k}\right) + \left(\varphi_0(x),\ \int_{-1}^{1} \frac{K(t,\ x)}{t^p} z(t)dt\right)\\ &\quad + \sum_{k=1}^{p-1} \alpha_k \left(\frac{1}{x^k},\ \int_{-1}^{1} \frac{K(t,\ x)}{t^p} z(t)dt\right) + \sum_{n=0}^{p-1} \beta_n (-1)^n \left(K_1^{\{n\}}(0,\ x),\ \varphi_0(x)\right)\\ &\quad + \sum_{k=1}^{p-1} \alpha_n \sum_{n=0}^{p-1} (-1)^n \beta_n \left(K_1^{\{n\}}(0,\ x),\ \frac{1}{x^k}\right).\end{aligned} \tag{35}$$

Now, we are able to compare $\left(\hat{A}\varphi,\ \Psi\right)$ and $\left(\varphi,\ \hat{A}'\Psi\right)$. Therefore, we obtain the equality between the terms considered for every $\varphi(x) \in V_m$ and for every $\Psi \in P^1$, only if and only if it is taking place in the following relationship:

$$\sum_{k=1}^{p-1} \alpha_k \left(z'(x),\ \frac{1}{x^k}\right) = \sum_{k=1}^{p-1} (-1)^{k-1} \alpha_{p-k}\beta_{k-1}(p-k)(k-1)! + \sum_{k=1}^{p-1} k\alpha_k \int_{-1}^{1} \frac{z(x)}{x^{k+1}} dx, \tag{36}$$

where β_{p-1} is an arbitrary constant.

In other words, the operators $\hat{A}$ and $\hat{A}'$ are associated operators only if and only, when it is accomplished under the following conditions:

$$\alpha_k \int_{-1}^{1} \frac{z'(x)}{x^k} dx = (-1)^{k-1} \alpha_{p-k}\beta_{k-1}(p-k)(k-1)! + k\alpha_k \int_{-1}^{1} \frac{z(x)}{x^{k+1}} dx \tag{37}$$

for every $k = 1, \dots, p-1$.

From condition (37) we can express the parameters β_{k-1} through the function $z(x)$, that is going to give us:

$$\beta_{k-1} = \frac{(-1)^{k-1}}{(k-1)!(p-k)\alpha_{p-k}} \alpha_k \left(\int_{-1}^{1} \frac{z'(x)}{x^k} dx - k\int_{-1}^{1} \frac{z(x)}{x^{k+1}} dx\right),\quad k = 1, \dots, p-1. \tag{38}$$

Therefore, if we note by $\hat{P^1}$ the restriction of the space P^1 with the condition (37), then it has the following form:

$$\hat{P^1} = \left\{ \Psi(x) \in P^1 / \Psi(x) = \frac{z(x)}{x^p} + \sum_{k=0}^{p-2} \beta_k \delta^{\{k\}}(x) + \beta_{p-1}\delta^{\{p-1\}}(x) \right\}, \tag{39}$$

where β_k, $k = 0, \ldots, p-2$ are defined by the formula (38) and β_{p-1} is an arbitrary constant.

The restriction of the operator A' on the space $\hat{P^1}$ we denote by the following way $\hat{A}'$.

Next, we note $\hat{A}$ as previously the extension of the operator A on V_m. Then the following operators:

$\hat{A}: V_m \rightarrow C_0^{\{P\}}[-1, 1]$ and $\hat{A}' : \hat{P^1} \rightarrow C_0^{\{P\}}[-1, 1]$ on the basis of previously done computations are verifying the established relationship:

$$\left(\hat{A}\varphi, \Psi\right) = \left(\varphi, \hat{A}'\Psi\right) \tag{40}$$

for every $\varphi(x) \in V_m$ and for every $\Psi \in \hat{P^1}$, so that they are associated operators.

As operator $\hat{A}$ is the extension of the operator A on $(p-1)-$ dimensional space, then the operator $\hat{A}: V_m \rightarrow C_0^{\{P\}}[-1, 1]$ is a noether operator with the index $\chi\left(\hat{A}\right) = -p + (p-1) = -1$.

Next, as the operator $\hat{A}'$ is the restriction of the operator A′ on $(p-1)-$ conditions (38), then the operator $\hat{A}' : \hat{P^1} \rightarrow C_0^{\{P\}}[-1, 1]$ is also a noether operator and its index $\chi\left(\hat{A}'\right) = p - (p-1) = 1$.

All that has been said allow us to formulate the result on noetherity of the extended operator $\hat{A}$, that is what is given by virtue of Duduchava's Lemma this following important global theorem:

Theorem 3.1. The equation $\hat{A}\,\varphi = f$, where $\hat{A}$ is the extended operator of the operator A of the form (1) and $f(x) \in C_0^{\{P\}}[-1, 1]$ is solvable in the space V_m only if and only when $\int_{-1}^{1} f(t)\Psi_k(t)dt = 0, k = 1,2,3, \ldots, \alpha\left(\hat{A}'\right)$, where $\{\Psi_k\}-$ is the basis of solutions of the associated homogeneous equation $\hat{A}'\Psi = 0$ in the associated space $\hat{P^1}$.

Before concluding, let us make an important remark.

Remark

The requirements (38) allow us to write in a more clear way the form of the functions from the associated space $\hat{P^1}$, i.e.,

$$\Psi(x) \in \hat{P^1} \Leftrightarrow \Psi(x) = \frac{z(x)}{x^p} + \sum_{k=1}^{p-1} \frac{(-1)^{k-1}}{(k-1)!(p-k)\alpha_{p-k}} \alpha_k \left(\int_{-1}^{1} \frac{z'(x)}{x^k} dx - k \int_{-1}^{1} \frac{z(x)}{x^{k+1}} dx \right) \delta^{\{k-1\}}(x) + \beta_{k-1} \delta^{\{p-1\}}(x), \tag{41}$$

where β_{p-1} is an arbitrary constant.

4. Conclusion

Summarizing our work, we state that we have completely realized the extension of a noether operator A defined by the extended operator $\hat{A}$ in the space V_m. We applied the well-known noether theory for integrodifferential operators defined by a third kind integral equation and, we computed very attentively both the two expressions $\left(\hat{A}\varphi, \Psi\right)$ and $\left(\varphi, \hat{A}'\Psi\right)$, taking the functions φ and Ψ respectively from the generalized functional spaces V_m and $\hat{P^1}$. From the previous, we released the conditions under which the two operators $\hat{A}$ and $\hat{A}'$ are associated operators. Consequently, we formalized within theorem 3.1 the global results of the investigation related to the question of noetherity nature of the extended operator, which as proved is noether operator. The principle of the conservation of noetherity nature of a noether operator after extension by some finite dimensional space of added functions to the initial space is established firmly as proved in theory. We can also note that the index of the initial operator after extension remains the same, i.e., $\chi(A) = \chi\left(\hat{A}\right)$ no matter the deficient numbers may be other than the previous before extension, i.e., $(\alpha(A), \beta(A)) \neq \left(\alpha\left(\hat{A}\right), \beta\left(\hat{A}\right)\right)$.

5. Recommendations

The achieved researches in this work completed by those already obtained by many scientific researchers related to the question of the noetherity nature of an extended operator of an initial noether operator in some various functional generalized spaces may lead us to project, and to set very interesting and challenging future investigations for noetherity when at this time, we take the unknown generalized function from the space $T_m = C^1_{-1}[-1, 1] \oplus \left\{\sum_{k=0}^{m} \alpha_k \delta^{\{k\}}(x)\right\} \oplus \left\{\sum_{k=1}^{m} \alpha_k F.p \frac{1}{x^k}\right\}$, as previously done by many scientists in their researchers, namely cited Gobbassov N. S, Raslambekov S. N, Bart G. R, and Warnock R. L.

This will be the next work to be done in a brief future. We underline once more again that the main difficulty appearing when realizing such extension is still and always connected with the derivative of the unknown function within the third kind singular integral equation through which is defined the initial integrodifferential operator to be extended onto the new generalized functional space.

Acknowledgements

The author is very grateful to the late Professor KARAPETYANTS N.K former Head of the Department of Differential and Integral Equations of the Rostov State University in the Russian Federation, for his invaluable support and helpful discussions during that time when conducting such research.

Author details

Abdourahman Haman Adji
Department of Mathematics, Higher Teachers' Training College, University of Maroua, Maroua, Cameroon

*Address all correspondence to: abdoulshehou@yahoo.fr

References

[1] Shulaia D. Solution of a linear integral equation of third kind. Georgian Mathematical Journal. 2002;**9**: 179-196

[2] Shulaia D. Integral equations of third kind for the case of piecewise monotone coefficients. Transactions of A. Razmadze Math. Institute. 2017;**171**:396-410

[3] Rogozhin VS, Raslambekov SN. Noether theory of integral equations of the third kind in the space of continuous and generalized functions. Soviet Math. (Iz. VUZ). 1979;**23**(1):48-53

[4] Abdourahman A. On a Linear Integral Equation of the Third Kind with a Singular Differential Operator Inthe Main Part. Moscow: Rostov-na-Donu, deposited in VINITI; 2002. pp. 560-B2002

[5] Abdourahman A, Karapetiants N. Noether theory for third kind linear integral equation with a singular linear differential operator in the main part. Proceedings of A. Razmadze Math. Institute. 2004;**135**:1-26

[6] G.I. Vinogradova. Singular integral operators in the axis with exact linear shift in space with weight, Math. Investigations. 1979;**3**:67–72. (in Russian)

[7] Bart GR, Warnock RL. Solutions of a nonlinear integral equation for high energy scattering. III. Analyticity of solutions in a parameter explored numerically. Journal of Mathematical Physics. 1972;**13**:1896-1902

[8] Bart GR, Johnson PW, Warnock RL. Continuum ambiguity in the construction of unitary analytic amplitudes from fixed-energy scattering data. Journal of Mathematical Physics. 1973;**14**:1558-1565

[9] Rogozhin VS. Noether Theory of Operators. 2nd ed. Vol. 99. Rostov-na-Donu: Izdat. Rostov Univ; 1982. p. 23

[10] Abdourahman. On a linear integral equation of the third kind with singularity in the main part. In: Abstract of Reports. International School-Seminar in Geometry and Analysis Dedicated to the 90th Year N.V Efimov. Abrao-Diurso: Rostov State University; 2000. pp. 86-87

[11] Duduchava R.V. Singular integral equations in the Holder spaces with weight.I.Holder coefficients. Mathematics Researches.T.V,2nd Edition.(1970)Pp104−124.Volume:5, Issue:2.ISSN:0542−9994

[12] Abdourahman ET, Weimbapou EK. Noetherity of a Dirac Delta-extension for a Noether operator. International Journal of Theoretical and Applied Mathematics. 2022;**8**(3):51-57. DOI: 10.11648/j.ijtam.20220803.11

[13] Karapetiants NS, Samko SG. Equations with Involutive Operators. Boston–Basel–Berlin: Birkhauser; 2001

[14] Prossdorf S. Some Classes of Singular Equations. Moscow: Mir; 1979 [in Russian]

[15] Gobbasov NS. A special version of the collocation method for integral equations of the third kind. Diff. Equ. 2005;**41**:1768-1774

[16] Gobbasov NS. Metody Resheniya integral'nykh Uravnenii Fredgol'ma v Prostranstvakh Obobshchennykh Funktsii (Methods for Solving Fredholm

Integral Equations in Spaces of Distributions). Un-ta, Kazan: Izd-vo Kazan; 2006 [in Russian]

[17] Yurko V.A. On a differential operators of higher order with singularities inside the interval. Kratkie sochenie Mathematicheskie Zamietkie, 2002. T. 71, No. 1, P. 152-156

[18] Gohberg I, Krupnik N. Extension theorems for Fredholm and invertibility symbols. In: Operator Theory: Advances and Applications. Vol. 16. Basel: Birkhäuser Verlag; 1993. pp. 514-529

[19] Ferziger JH, Kaper HG. Mathematical Theory of Transport Processes in Gases. Amsterdam–London: North-Holland Publ. Company; 1972

[20] Hilbert D. Grundzüge einer allgemeinen Theorie der linear Integralgleichungen. New York: Chelsea Publ. Company; 1953

[21] Picard E. Un théorème général sur certaines équations intégrales de troisième espèce. Comptes Rendus. 1910; **150**:489-491

[22] Weimbapou ET, Abdourahman, Kengne E. On Delta-extension for a Noether operator. Russian Mathematics. 2021;**65**(11):34-45 Allerton Press, Inc., 2021. Russian Text The Author(s), 2021, published in Izvestiya Vysshikh Uchebnykh Zavedenii. Matematika, 2021, No. 11, pp. 40–53

[23] Terbeche M, Sacaliuc G, Vornicescu G. Solutions of some singular integral equations. Bulletindes Sciences Mathématiques. 2000;**126**:379-387

[24] Krupnik N, Nyaga VI. Onsingular integral operators with shift in the case of piecewise Lyapunov contour. Notice of AN Georgia SSR. 1974;**76**(1): 25-28. (in Russian)

[25] Rogozhin VS, RNS. The Noether theory for integral equations of the third kind. Differentsial'nye Uravneniya. 1978;**14**(9):1678-1686 Math-Net. Ru All Russian Mathematical Portal

[26] Yurko VA. Integral transforms connected with differential operators having singularities inside the interval. Integral Transforms and Special Functions. 1997;**5**(3–4):309-322

[27] Bart GR. Three theorems on third kind linear integral equations. Journal of Mathematical Analysis and Applications. 1981;**79**:48-57

[28] Bart GR, Warnock RL. Linear integral equations of the third kind. SIAM Journal on Mathematical Analysis. 1973;**4**:609-622

[29] Sukavanam N. A Fredholm-type theory for third kind linear integral equations. J. Math. Analysis Appl. 1984; **100**:478-484

[30] Shulaia D. On one Fredholm integral equation of third kind. Georgian Mathematical Journal. 1997;**4**:464-476

[31] Gobbasov NS. Methods for solving an integral equation of the third kind with fixed singularities in the kernel. Diff. Equ. 2009;**45**:1341-1348